COUNTY DURHAM
Place-Names

Published by Sigma Leisure – an imprint of Sigma Press, Stobart House, Pontyclerc, Penybanc Road, Ammanford, Carmarthenshire SA18 3HP.

British Library Cataloguing in Publication Data
A CIP record for this book is available from the British Library.

ISBN: 978-1-85058-995-2

Typesetting and Design by: Sigma Press, Ammanford, Carmarthenshire

Cover photographs: © Anthony Poulton-Smith
main picture: Imposing gateway at Auckland Castle
top (left to right): Colliery public house sign, Crook; An unusual sign at Edmondsley; Killerby sign; Pity Me sign

Photographs: © Anthony Poulton-Smith

Printed by: TJ International Ltd, Padstow, Cornwall

Disclaimer: the information in this book is given in good faith and is believed to be correct at the time of publication. No responsibility is accepted by either the author or publisher for errors or omissions.

COUNTY DURHAM
Place-Names

Anthony Poulton-Smith

Contents

Introduction

For years the history of England was based on the Roman occupation. In recent years we have come to realise the influence of the Empire did not completely rewrite British history, indeed there was already a thriving culture in England well before the birth of Christ. When the Romans left our shores in the fifth century the arrival of the Anglo-Saxons was thought to herald a time of turmoil, yet they brought the culture and language which forms the basis of modern England. Later the arrival of the Norsemen saw their influence and the same is true of our place names, the vast majority of settlement names in County Durham are derived from the Saxon/Old English or Old Scandinavian tongues. There are also the topographical features such as rivers and hills still have names given to them by the Celts of the pre-Roman era.

Ostensibly place names are simply descriptions of the location, or of the uses and the people who lived there. In the pages that follow an examination of the origins and meanings of the names in County Durham will reveal all. Not only will we see Saxon and Scandinavian settlements, but Celtic rivers and hills, Roman roads and even Norman French landlords who have all contributed to the evolution to some degree to the names we are otherwise so familiar with.

Not only are the basic names discussed but also districts, hills, streams, fields, roads, lanes, streets and public houses. Road and street names are normally of more recent derivation, named after those who played a significant role in the development of a town or revealing what existed in the village before the developers moved in. The benefactors who provided housing and employment in the eighteenth and nineteenth centuries are often forgotten, yet their names live on in the name found on the sign at the end of the street and often have a story to tell. Pub names are almost a language of their own. Again they are not named arbitrarily but are based on the history of the place and can open a new window on the history of our towns and villages.

Defining place names of all varieties can give an insight into history which would otherwise be ignored or even lost. In the ensuing pages we shall examine 2,000 plus years of Durham history. While driving around this area the author was delighted by the unique place names found in the county and so, having already taken a look at, among others, *Cambridgeshire Place Names*, turned here to County Durham. This book is the result of the author's long interest in place names which has developed over many years and is the latest in a series which continues to intrigue and surprise.

To all who helped in my research, from the librarians who produced the written word to those who pointed a lost traveller in the right direction, a big thank you.

Place-Names

(in alphabetic order)

A

Annfield Plain

A village of some three thousand today, this had never been particularly populous before the coming of the railways and coal mining to the area. However this is not to say the name is not much older, indeed this almost certainly represents Old English *feld* and a Saxon personal name and gives 'the open land of a man called Ann'.

The additional 'Plain' is very recent, not seen before the nineteenth century, even so it has managed change in this short period of time. Here the origin is not the flatland 'plain' but the term used to refer to the gentle slope known as Annfield Plane and only appearing on maps after the coming of the railway transporting coal and limestone. Initially powered by the engine housed on nearby Loud Mill, it became a part of the national network when linked to the Stockton and Darlington line in 1845, by which time the Derwent Iron Company were also using it. For once we know exactly when the name changed, indeed it was a quite deliberate change in 1856 when the first cottages were built to house the miners.

Aycliffe

Records begin with a document dated around 1085, when it appears as Aclea. Clearly this is Old English *ac leah* and describes 'the oak trees at the woodland clearing'.

Middridge is a local name found as Midderrigg, Midrige, Midrich, Mitrich, Midregg, and Midderigge and describes 'the middle ridge of land'. What is today given as Nunstainton was Nun Staynton, the 'farmstead on stony ground' and land given by

Fingerpost near Middridge

Iveta of Arches to her sister, who was prioress of the convent here. Locally the name of Ricknall survives most obviously in Ricknall Lane. It is derived from a Saxon personal name and Old English *healh* and describes 'the nook of land of a man called Rica'. Woodham is a common place name, referring to 'the wood at the hemmed in land'. Nunstainton is recorded as Staynton supra Schyram in 1190, this being 'the stony farmstead on the River Skerne'. Later we find the addition of the prefix *nunenna* referring possession by Monkton.

The Bay Horse public house, Middridge

Heworth is from Old English *heah worth* 'the high enclosure'. Ketton is seen as Cathona in 1091 and as Cattun in 1125, which seems to contain Middle English *ket* (also seen in Old Scandinavian) and meaning 'flesh'. Hence this is either *ket tun* or, and this seems more likely, this is used as a nickname and gives 'the farmstead of a man known as Kett'. The name of Woodham is easy to see as 'the homestead near the wood'.

The sign clearly depicts the attributes of the horse which gave its name to the Bay Horse at Middridge

The North Briton is a pub which undoubtedly was chosen to describe those who lived in the north of the land. However it could well have been influenced by the London pub of this name, itself from the periodical established by the politician John Wilkes which issued a string of attacks on his fellow members of parliament, the monarchy, and just about every corner of the establishment.

B

Barmpton

A small village on the banks of the River Skerne, this name is found as Bermetun in 1199, later entries as Bermeston, Bermentun, Bermpton, Barmtone and Barmeton. Here a Saxon personal name and Old English *tun* tells us this was 'the farmstead of a man called Beorn'.

Barnard Castle

Listed as Castellum Bernardi in 1200, this name tells us of 'the castle of a man called Bernard'. The man is well documented as being here in the twelfth century. To those who live hereabouts the place is known simply as Barney.

The castle was built during the Norman era of the early eleventh century by Bernard Balliol. His father, Guy de Baliol, came to England with William the Conqueror and was given the land here by William Rufus in 1093. As the castle provided local employment it was natural for a community to cluster and grow around it. During the sixteenth century the Rising of the North saw the castle besieged as supporters of Mary, Queen of Scots surrounded it. Sir George Bowes managed to hold out for eleven days before surrendering. The Balliol family name is remembered in the name of Balliol Street, while the local museum is named after the Bowes family and built in the late nineteenth century by John and Josephine Bowes.

To the north we find the hamlet of Kinninvie, a place which has grown solely because of its position at a minor crossroads. Here the first element is clearly a personal name, however the suffix is less obvious for early forms are unknown. It is supposed this represents a corruption of Old Scandinavian *by*, in which case this would be 'the farmstead of a man called Cinna or Cynna'.

Marwood is a local name from *maer wudu* 'the boundary wood'. Shotton is derived from *scot dun*, Old English for 'the hill of the huts'. The Sills is a shortened form of the fourteenth century record of Suleshop, itself describing 'the small valley of a man called Syla' and the name seen as Sils by 1723. Streatlam is easily seen as from Old English *straet leah* 'the woodland clearing by a Roman road'.

The Beaconsfield is a local pub named after former British prime minister Benjamin Disraeli, who retired to the House of Lords as the Earl of Beaconsfield. The Three Horse Shoes is a reminder of the days when the village blacksmith and the public house were found alongside one another, providing the same service as the modern motorway service station. Indeed the Three Horse Shoes should be seen as a question, offering the traveller a replacement shoe for his mount. The Blacksmiths Arms has identical origins.

The Turks Head is heraldic, the image showing a member of the family fought on the side of the Christians in one of the Crusades to the Holy Land. Another patriotic pub name is the George and Dragon Inn, recalling the English patron saint and his most famous exploit. Clearly the Old Well is a pub which points to an early water source being found nearby. The Moorcock Inn probably shows the alternative name for the red grouse, although that would normally be seen as two words. The horse called Moorcock which won the Richmond Gold Cup in three successive years cannot entirely be ruled out. The Red Well Inn takes the name of a place name describing the 'spring where reeds grow'.

Barningham

Domesday records this name as Berningham. Here is a Saxon personal name and Old English *inga ham* and referring to 'the homestead of the family or followers of a man called Beorn'.

Beamish

Today the place is best known for its excellent open air museum. In the thirteenth century the place was famous for its 'beautiful mansion', the name coming from Old French *beau mes* and listed as Bewmys in 1288.

Pub names include the Shepherd and Shepherdess, traditionally held to be two metal figures not seen since the Napoleonic Wars when they were given to be melted down to be made into ammunition. Beamish Mary was named from the local colliery and is but one of several rather unusual names found here. The local church is known as the Tin Chapel, a clear reference to the amount of metal used in putting up this building in double quick time.

However surely one of the oddest place names in the country is the village of No Place. The origin is something of a mystery, several suggestions have been put forward, such as corruptions of Nigh Place or North Place, but neither are really likely. There are several places in the country named for not being identified as belonging to any of its neighbours and this is the most plausible explanation. However it is by no means certain as the name went unrecorded until the building of just four terraced houses which became known as No Place Villas. These properties were demolished in 1937 and the name applied to an area just to the north where another group of houses had been built known as Co-operative Villas. In 1983 it was decided to change the name of the village to Co-operative Villas but this met with strong opposition and now road signs point to No Place and to Co-operative Villas, even though both are the same place.

Bearpark

As with the previous entry this comes from Old French. Here the origin is *beau repaire*, is listed as Beaurepayre in 1267, and speaks of 'the beautiful retreat'.

Minor names of the parish include Alden Grange, the name is found as Aldyngrigge in 1416, with later fifteenth century records of Aldynrige and Aldyngraunge. The modern name takes the Saxon personal name Ealda and adds another reference to the area being a hunting retreat. Note this was not always the case, for the earlier two fifteenth century forms have the Old English suffix *hrycg* giving 'the ridge of land of a man called Aslac'.

Stotgate features Old English *stott geat*, the second element is the origin of the modern 'gate' but in Saxon times referred to the 'way' and not a barrier across it. Similarly while Stott is a personal name, it is also a dialect word referring to 'a heifer'. Together this could well be suggesting a drover's route to market. Despite first being recorded several centuries ago, Lodge Hill still explains itself.

Bedburn

Recorded as Bedburn in 1291, it comes from a Saxon personal name and Old English *burna* and describing 'the stream of a man called Beda'.

The minor name of Podge Hole is found as Poydeshole in a document dated 1382. While this seems to point to 'the hollow of a man called Poid', it must be noted how Old Scandinavian *poid* describes 'a vile person'. It seems unlikely any parents would refer to their son in such a way, hence it may have been used as a nickname by the neighbours who effectively named the place.

Shipley, found as Shepley in 1349 and in the modern form for the first time in 1382, is a common name in the north of England coming from *sceap leah* 'the woodland clearing where sheep are raised'. Rackwood features a Saxon personal name and Old English *wudu* giving 'the wood of a man called Hraca'. Crowsfield is quite late, first seen in 1491, and refers to 'the *feld* or open land of the family of Robert Crawe'.

The first element of Chatterley is an uncertain personal name, possibly something akin to 'the *leah* or woodland clearing of a man called Cadhere'. Euden Beck is found as Edeneburn in 1311 and Euedenburn in 1441, this coming from Old English for 'the stream in the valley of yew trees'. Harperley is seen as Harperleia in 1382, this being 'the woodland clearing of a man called Harper'. Wadley comes from Old English *wad leah* and describes 'the woodland clearing where woad grows'. Woodfield describes 'the *feld* or open land of a man called Wuda'.

Belmont

A village which has now been largely swallowed up by the encroaching city of Durham, it does still rank as a parish and still tries to maintain its own identity. The name is simple enough, for there is no doubt this represents Old French *belle mont* 'the beautiful hill'.

Benfieldside

While this is a parish it is not the name of a town or a village. Seen as Benelands and Benefeldside in the twelfth century, it seems the original name spoke of 'the cultivated land where beans are grown' while the later form seems to be referring to the road which ran alongside 'the open land where beans are grown'.

Billingham

Records of this name include Billingham in 800, Billingeham in 1291. This is from Old English where the Saxon personal name is suffixed by *ing ham* and refers to 'the homestead of the family or followers of Billa'.

Locally we find names such as Belasis Hall, which comes from Old French *bel assise* and describes 'the beautiful site'. Haverton Hill is derived from *hafer dun* and describes 'a humped hill'. Port Clarence is a name derived from the Clarence Railway which connected to the Stockton and Darlington at Sim Pasture, in turn this comes from the third son of George III, William Duke of Clarence ascended to the throne in 1830 and became William IV.

Field names here are diverse and include Bone Orchard, named after John Bone who was resident here in 1609. Brandreth Head quite literally describes 'a tripod or framework on which a hayrick is made', almost certainly pointing to the site of a beacon. Flutter Carr is derived from *flodor ker* and referring to 'a boggy place liable to flood during very wet times'.

Bellasis is a minor name found as Belsis, Belleys, Bellacyse, Bellais, Bellace and Bellasez over the years. This is probably Old French *bell assize* or 'the good or reliable place with a fixed rent', although the first element might be the Saxon personal name Belle. Haverton Hill is hardly the steepest of inclines nor is it the highest of summits, however this is purely comparative for this place near the mouth of the Tees most likely represents *haeford tun* and 'the farmstead of or by the headland'. Bewley puts together Old French *beau* and Old English *leah* and speaks of 'the beautiful woodland clearing'. Norton is among the most common of names, always describing 'the northern farmstead'.

Birtley

Recorded as Britlei, Brittele and Byrtelay in the thirteenth century, and as Bircteley, Brithley and Bireteley in the fourteenth, this name has two possible Old English origins. Most likely this represents *beorht leah* describing 'the bright woodland clearing', although the first element may well represent a Saxon personal name.

Ouston is found as Ulkestan in 1328 and Ulleston in 1382, which point to an origin of 'the farmstead of a man called Ulla'. The only known record of Sheddon's Hill dates from 1382 as Scedneslawe, here a Saxon personal name precedes Old English *hlaw* and is likely to come from 'the hill of a man called Sceldwine'.

Buteland is found as the modern form as early as 1255, this is 'the farmland of a man called Bota'. Slaggyford is not a reference to waste from smelting, this is a dialect *slag* more often recorded in Scotland and here describing 'the muddy ford'. Tone is recorded as Tolland in 1182 and again in 1296, this is probably from Old English *toll land* or 'the agricultural land on which a toll is due'.

One local is the Moulders Arms, which shows a link with those who made the moulds for iron-casting, probably a former landlord or owner. The Barley Mow recalls the stack of barley, a vital ingredient in brewing and an instantly recognised as an establishment brewing its own beer. The Vigo Inn is named after the Battle of Vigo Bay, where the combined British and Dutch fleets destroyed the Spanish ships in 1702. The Bowes Incline Hotel is a reminder of mining, George Stephenson's 1826 standard gauge cable railway system is still operational and the only example of such in the world. Alongside the car park visitors can still see one of the original wagons.

Bishop Auckland

A name listed as Alclit in a document dated around 1040. This is probably a Celtic name describing 'the rock or hill on a river called Clyde'. Clearly these is no river here called Clyde today, this referring to 'the cleansing one', thus we must assume this was an earlier name of the River Gaunless, itself from Old Scandinavian *gagnlauss* and describing 'the unprofitable one'.

The addition here comes from possession by the Bishop of Durham.

Eldon is found as Heldun in 1204, this comes from Old English and describes 'the hill of a man called Ealla'. Next door is Old Eldon, a name of similar origins. Etherley is seen as Etherdacres in the fourteenth century, this began as 'the open land of a man called Aethred'. Bracks Farm has been mistaken for a personal name, although it may well have been used as such in later years it began as *braec* or 'newly broken land for cultivation'.

Attractive sign for Eldon

Bondgate is an historic street showing it occupied land owned by the bondsmen of the town.

Coundon is from Old English *cuna dun* 'the hill where cows are reared'. Early records of Farnley, as Le Farmley in 1313 and as Farmley in 1399, show the first element here to be corrupt and this is actually 'the woodland clearing of the farm'. This is not 'a farm' in the

Old Eldon's modern sign

modern sense but refers to land held at a fixed rent. Henknowle, still sometimes seen as the earlier form of Henknoll, comes from 'the *cnoll* or hill of a man called Hean'.

Remnants of the estate of the lord of the manor are seen in the Lodge, now a popular local pub. A name such as the Beehive does not mean a hive was nearby, the name most often taken for the simplicity of its shape, although the association of 'busy bees' and 'a hive of industry' may also have influenced the decision. The Hare and Hounds instantly makes us think of hare coursing, this blood sport (where the hare is rarely caught) may not be the true origin, for alliteration is common to pub names and the name could have been taken for either reason equally.

Early records of Fielden Bridge include Feldyngford and Fyldynngate at either end of the fourteenth century. Clearly the bridge replaced the ford, the *geat* simply another reference to the 'way' or 'road' across the river. The first element is less certain, it certainly seems to be 'Fielding', yet this is too early to be a surname and 'the dwellers of the open land' is highly unlikely. Beechburn is a little misleading for the first element is not a tree but a personal name, this is 'the stream associated with a man called Bicca'. Fitches has proven difficult to define, not least because so few early forms exist, those limited to Fychewache in 1382 and Fyccheworth in 1392, which be *fiche worth* 'the enclosure marked by vetch'. Henknowle speaks of 'the hillock of a man called Henna'.

Although he was born in Lancashire, Stan Laurel spent his early years in Bishop Auckland, the Comedian public house being named to mark his early years here. It was inevitable that at least one pub would be named the Weardale Inn, the name given to the area through which the river flows. Brewing is an obvious theme, the tools include the wooden malting shovel, easily recognised by large blade. Some of the earliest pubs known as the Malt Shovel would have used an old shovel or possibly a replica as the sign. The Punch Bowl Inn is another example of the product being advertised, yet such was ever seen here other than the image on the sign. Tut 'n' Shive are tools used for tapping a barrel, the container made by those mention by the Coopers Arms.

The Black Bull Inn is one of the many names which is probably from an unknown coat of arms, although there must be some examples named after a favourite animal. The Mill Race refers to the channel dug to provide a good flow of water over the mill wheel. The original Stags Head probably featured an actual head as its sign, or at least the antlers, most will refer to the hunt. No name can offer a warmer invitation than the Welcome, while no doubt exists of the former use of the Station public house. Outside the Cumberland Arms is the coat of arms of the Clifford family, Earls of Cumberland. While the Ship Inn probably refers to the ocean, it should be realised that at least half these pubs, and particularly those found away from the coast, feature a corruption of *sceap* or 'sheep'.

Of course horses have four legs, thus the name of the Three Horseshoes seems odd until it is viewed as a question. If the answer to that question is affirmative, the solution lies alongside the pub in the form of the blacksmith, who takes care of the horse while the inn

above: Imposing gateway at Auckland Castle
right: The grand Town Hall building at Bishop Auckland

Church of St Andrew Auckland, South Church

entertains the rider. The Trotters Inn shows a trotting pony, a later mode of transport. The Eden Arms was home to the Eden family. George Eden, Earl of Auckland (1784-1849), was a Whig politician who served three times as First Lord of the Admiralty and as Governor-General of India. The local theatre is also named after this influential family. From the refrain "Ha, ha, ha, he, he, he, little brown jug, don't I love thee" comes the inspiration for the Brown Jug, which suggests a drink of ale, a pint glass with a handle referred to as 'a jug'.

Bishop Middleham
Recorded as Middelham in the twelfth century, this comes from Old English *middel ham* and comes as no surprise to find this means 'the middle homestead'. The addition, to be expected when the name is fairly common, refers to the early possession by the Bishop of Durham.

Cornforth comes from *corna ford* 'the ford of the cranes or herons'. From Old English *thynne ford* comes Thinford or 'the ford which is not densely overgrown'. Stob Cross Lane takes its name from *stob crouche* 'the cross marked by a stump'. Stotforth speaks of 'the fold for the domesticated animals', from *stot fold*. What was once known as 'the ford of a man called Maegen' is now seen as Mainsforth. Thirslington comes from 'the farmstead associated with a man called Thurstan'.

Garmondsway started out as 'the road of a man called Garmund'. Island Farm should not be taken literally, however there was a single acre of land here, farmed quite separately from the rest. Island Farm is from Old English *eg land* meaning 'the cultivated land on the dry ground in a marsh'.

The Cross Keys Inn is another pub which shows the link between the pub and the church. The parish church is dedicated to St Michael, while the public house is named after the symbol for St Peter. Ye Olde Fleece Inn is clearly not that old, no matter what the name says. The term 'Ye Olde' may be synonymous with age but is quite the reverse, for it originates in the US and came to our shores in the late nineteenth or early twentieth centuries. The Fleece tells us it was associated with the wool trade, fundamentally driving the British economy and making the country the global power from the late thirteenth century. The Dun Cow is a description of its colour, a light brown, which would certainly refer to a favourite local animal.

As noted earlier, the name is a reference to the Bishop of Durham. In 1883 the *Boldon Book*, a survey of the lands held by the Bishop of Durham, shows there were 32 houses here. Details of this survey, when compared to Domesday, are highly detailed, even to the point where individuals are mentioned. The village was a favourite of several bishops, two even died here, although their residence can only be seen in the form of an earthwork.

The village was on an important route during the Roman era, still known as Cades Road. However no Roman buildings have yet been discovered, although artefacts have been recovered. A bronze statue of a Roman god, and four pans decorated as to suggest they were used in a religious rite probably indicate the village once had a Roman temple. To date no evidence of this temple has been found but surely can only be a matter of time.

Bishopton

Found in a document from 1104 as Biscoptun, this name is derived from Old English *biscop tun* and tells us of 'the farmstead of the bishop'.

Here, too, are the names of Castle Hill, a pointer to the ancient mound which still casts a shadow over the parish; Gilly Flat is 'the even ground of a man called Giles'; Stone Flat tells of 'the stony even ground'; predictably horses were exercised on Galloping Hills; and the oddly named Woogra comes from *wulf grava* and describes 'the grove frequented by wolves'.

Blackwell

Here is a name which explains itself. From Old English *blaec wella*, and recorded as Blackewell in 964 and as Blachewelle in 1086, this describes 'the dark coloured stream'.

Blaydon

Records of this name include Blakden in 1333, Blakedene in 1345 and Bladal in the early sixteenth century. This is derived from a Saxon personal name and Old English *denu* to refer to 'the valley of a man called Blac'.

Locally we find the hamlet of Barlow, found in the twelfth century as Berley and Berlei. While the suffix here would normally be said to be *hlaw* here the two Old English elements are clearly *bere leah* and tells of 'the open land where barley is grown'.

The growth of Blaydon was rapid following the siting of a smelting operation here at the end of the eighteenth century. Prior to the establishment of industry at what should correctly be called Blaydon-on-Tyne, although hardly ever used outside of official circles, was a small collection of farms and cottages.

Bolam

The earliest surviving record of this name comes from 1317 as Bolom. Here the name describes the '(place at) the tree trunks', this is a name which is derived from either Old English *bola* or Old Scandinavian *bolr*, the meaning is the same.

Trewick is a local name coming from Old English *treow wic*. Most often *wic* is defined as 'specialised farm' and almost every time that speciality was dairy farming. Here the name describes 'the dairy farm marked by a tree'. The rural location of Bolam is reflected in the name of its public house, the Countryman.

Boldon

Here is a name from Old English *bol dun* and the dwelling on a hill' or 'the hill of a man called Bol', should the first element be a personal name.

Minor names include Scots House, the name coming from the family of Galfridus Scot who held this manor. Strother comes from *strother*, an Old English term meaning 'the marshland overgrown with brushwood'.

Boldron

Around 1180 this name appears as Bolrum, this is derived from Old Scandinavian *boli rum* and refers to 'the clearing used by bulls'.

Bowes

Whether this originates from Old English *boga* or Old Scandinavian *bogi*, this refers to the '(place at) the river bends'. The name is recorded as Bogas in 1148.

The Ancient Unicorn Inn features the fabled creature which, this close to the border, must surely represent Scotland. As a general rule if a pub has to say it is 'old' in some way it almost certainly is not, the reference only added to suggest great age.

Brafferton

Listed as Bradfortuna in 1091, this name comes from Old English *brad ford tun* and tells us of 'the farmstead by the broad ford'.

Ketton comes from Old English, the Saxon personal name followed by *tun* and referring to 'the farmstead of a man called Ceatta'.

Brancepeth

Recorded as Brantespethe around 1170, this name comes from a Scandinavian personal name and Old English *paeth* and speaks of 'the road or way of a man called Brandr'.

Billy Row is found as Billyraw in 1425, this describing 'the hill of a man called Billing'. Burnigill is found as Burnynghill in 1313 and as Burnyngill in 1343, this being 'the hill associated with the family or followers of a man called Bruning'. Stockley is an Old English name from *stocc leah* 'the woodland clearing marked by stocks or tree stumps'. Crook is simple enough, it refers to 'the winding valley'.

Brancepeth Castle

Brandon

Found around 1190 as Bromedune, this comes from Old English *brom dun* and tells us it was known for being 'the hill where broom grows'.

Most residents of Brandon are probably grateful the place did not become known as Byshottles. There are two schools of thought as to the Old English origins of what is sometimes said to be Brandon's alternative name. Perhaps this is by *shodel* from *scaden* 'to divide', in which case this would be understood as 'the dividing road'. Alternatively this is by *scytels*, a name indicating the town's communal midden or latrine.

Brignall

Domesday lists this name as Bringhenale, which features a Saxon personal name and Old English *inga halh*. Together these speak of 'the nook of land of the family or followers of a man called Bryni'.

Broom

Found as Brom in 1170, this name comes from Old English *brom* and tells us it was the '(place) where broom grows'.

Aldin Grange is a local name found as Aldingrig in 1539, this being 'the ridge or edge of the family or followers of a man called Ealda'. Anton Field represents a personal name, nothing readily suggests itself from the records available but will probably have been someone named Anthony.

Browney (River)

Listed as Aqua de Brun in 1125 and as Brune flumen in 1170, the modern suffix is probably Old English *ea* 'water' and thus describing the colour of this river as 'the brown waters'.

Burdon

Records are many and include Birden, Burdon, Bireden, Burdun, Burton, and Birton. This is Old English *burh denu* and describes 'the stronghold in the valley'.

Burnhope

Found as Brunhop in 1317 and as Burnhop in 1382, this is 'the *hop* or small valley with a *burna* or stream'.

Locally we find Braeside, where the first element is from Old English *braew*, and related to Scots Gaelic *brae* and Welsh *bryn*, and even related to Sanskrit *bhru* 'eyebrow'. This refers to this being found at 'the side of brow of land'. Holmeside is a local name originating in 'the dry slope by the marsh'. Shorden Brae is seen as Schortedene in the thirteenth century, the name meaning 'the short valley'.

Burnopfield

Locals may tell you the village name comes from the time when border skirmishes with the Scots resulted in the inhabitants heroically setting fire to their own fields in order to stop the advance of the enemy. Literally this is said to mean 'burning up the fields'. The true meaning is much less romantic. Here Old English *burna hop feld* speaks of 'the open land by the stream in a valley'.

Minor names include Friarside, showing this was formerly denoted as monastic property. What began as 'Ibe's hillside' is marked on maps today as Gibside. The name of Leazes is a name most often seen as Leasowe from Old English *laeswe* and describing 'pasture lands'.

Butsfield

Also seen in the nearby East Butsfield, this name probably comes from a Saxon personal name and Old English *feld* to describe 'the open land of a man called Butt or Bud'. However the first element may be *butt* and here referring to that strip of land which remained unploughed at the edge of the field and where the plough team executed a turn. While this is normally only seen in field names, that this is found with *feld* may point to this origin. Note the Old English *feld* should not be seen as being exactly the same as the modern 'field' although it was certainly the forerunner. Today a field brings an image of a neatly hedged or fenced area, with access provided by a large five-barred gate. In Saxon times there was no closable gate and no perimeter hedge, other than the barrier made from the litter pushed to the sides when originally clearing the 'open land'.

Butterknowle

A name from Old English *buttere knoll* 'and referring to 'the small hill near the dairy farm'. Sometimes given as Butterknowle and Softley, the second name is also seen as a suburb, it must be from Middle English *soft* with Old English *leah* 'the soft woodland clearing' and probably should be seen as 'spongy'.

The local is known as the Diamond, not a reference to the precious stone but a name taken from Diamond Hill, named for its apparent shape.

Butterwick

From Old English *butere wic* and listed as Boterwyk in 1131, this is 'the specialised farm where butter is made'. Perhaps only one in every hundred specialist farms were known for something else for, as here, the vast majority specialised in dairy produce'.

Byers Green

Listed as Bires in 1345, this comes from Old English *byre* and describes 'the cowsheds', which are still known as byres in places.

Locally we find Binchester which, while the second element is certainly Old English *ceaster*, the first lacks sufficient records for us to be certain. However it seems the most likely origin is a Saxon personal name, thus giving 'the Roman stronghold of a man called Bini'. Wham is an Old English *hwamm* referring to 'the corner of land'.

C

Carlton

To the northwest of Stockton-on-Tees is this hamlet listed as Carlton, Careleton, and Carleton up to the thirteenth century. Here it is easy to see Old English *ceorl tun* referring to 'the farmstead of the free man'.

Cassop

Listed as Cazehope in 1183, this name comes from Old English *catt hop* and describes 'the valley frequented by wild cats'.

Castle Eden

A name first seen in 1050 when it was recorded as Iodene. This name comes from the Eden Burn, a Celtic river name meaning 'water' with Old English *burna* 'stream'. The 'castle' here is an eighteenth century mansion house.

The minor name of Hardwick is from Old English *heord wic* and documented as Herdwich, Herdewyk, Herdwic and Herdwyk. Normally the element *wic* would be said to be 'specialised farmstead, almost always dairy produce', yet here the first element confirms this is 'the farm for cattle'. Herrington is a Saxon settlement derived from 'the farmstead of the family or followers of Heor'. Hulam is a corruption of Old Scandinavian *holmr* 'the raised dry land in a marsh'.

Castleside

A name which points to the importance of the location and not a 'castle' we would know today. Here the crossroads of the main road between Darlington and Edinburgh and roads leading to Stanhope, Consett and the North Pennines. The 'castle' was simply a defensive feature.

Local names include Healey Field, a name first seen as Helaye and showing the second part of the name is comparatively recent and originally this was 'the *leah* or woodland clearing of a man called

Hael'. Rowley is a common Old English name from *ruh ley* and describes 'the rough woodland clearing'.

The locals enjoy a drink at the Smelters Arms. While metal workers are commonly seen in pub names, these come from an earlier age when travellers arrived by horse and later by coach. Smelting is the process of heating ore to extract the metal

Chester le Street

Records of this name include Cestra in 1160 and as Cestria in Strata in 1400. Here is a name from Old English *ceaster straet* and describing the 'Roman stronghold on the Roman road'. The later record features Latin *strata* with the same meaning, while the French definite article *le* is still seen despite the loss of the preposition.

Locally Breckon Hill is not recorded until comparatively recently, however the name is seen in Brakenburn and also in Brakenthwayth which refer to 'the undergrowth by the stream and in the woodland clearing' respectively and either could have given a name to the hill. Lambton is probably 'the farmstead where lambs are reared', although it is still possible the first element is a personal name even though none of the forms contain evidence of the possessive 's'.

Broomyholm is found as Bromywhome in 1326 and as Bromyngholm in 1382, this being 'the dry land in a marsh covered by broom'. The first element in Nettlesworth is probably the plant, hence 'the *worth* or enclosure overgrown with nettles', although a similar personal name cannot be ruled out entirely. Pelaw and Pelton share a Saxon personal name as the first element, in the case of the latter this is 'the *tun* or farmstead of a man called Pella'. However records of Pelaw sometimes suggest *hoh* 'the spur of land of a man called Pella' and others point to *hlaw* 'the mound of a man called Pella'.

Picktree, found as Pikeire and Piktre in the thirteenth century, is a little difficult but may refer to a 'pitch tree', one which produces much resin. Early records of Plawsworth make it difficult to see if this is *plaga worth* 'the enclosure near where play or sports takes place', or possibly 'the enclosure of a man called Plaw'. Rickleton is found as Rykelingden in 1339, this being 'the valley of the family or followers of a man called Ricel or Ricola'.

Cong Burn probably shares its first element with Coniscliffe, thus this is 'the stream of the king'. Harraton is listed as Hervertune in 1190, as Herverton in 1297, and as Harraton for the first time in 1562.

Lumley, recorded since the middle of the eleventh century, combines an Old Scandinavian personal name and Old English *leah* and speaks of 'the woodland clearing of a man called Lum or Lumi'. Urpeth has changed little since the earliest known form of Urpath in 1297. This is thought to be something akin to 'the path or way of a man called Ur'. While this is an unrecorded personal name, it would certainly be a pet form of names such as Urard, Uro, Urolf, or Urold.

Humour is a popular element in pub names, the Church Mouse being one example which hints at the poverty of the innkeeper. There is little doubt as to the association when it comes to the name of the Market Tavern. The Lambton Arms remembers John George Lambton, 1st Earl of Durham, who served as British ambassador to Russia, Austria, then later Germany, before becoming governor-general of the British holdings in North America. The Plough is one of the most common pub names, no wonder considering it represented the basic farming implement from a time when the vast majority toiled the fields. On Cookson Terrace is the Cookson, a public house traditionally held to be named after the author Catherine Cookson, however the pub takes the name of the road.

As a royal sport falconry was popular in Elizabethan times, hence the Falcon was almost an inevitable pub name. However it could well be heraldic, the image seen in the arms of Elizabeth I and Shakespeare and many others. The Rose and Shamrock may well be unique, it points to the union between England and Ireland.

The Olde Miners Lamp reminds us of the link with mining, however rather ironically the name tells us the place cannot be very old for the addition of 'Olde' or 'Ye Olde' is a very early twentieth century creation. The Warriors Arms is a general reference to fighting men, the Butchers Arms is a name dating back to times when the inn-keeper would double as the village butcher, the Smiths Arms reminds us the metalworker was invariably next door to the local.

Chilton

Found as Ciltona in 1091, this name is derived from Old English *cild tun* and describes 'the farm of the young noblemen'. This is a common place name, nearly always found with a second element for distinction.

Cleadon

The earliest records of this name include Clevedon and Clyvedon, showing this to be Old English *clif dun* and thus describing 'the hill with a bank or slope'.

Cleatlam

A name found as Cletlinga in 1050 and as Cletlum in 1271. The modern form is derived from Old English *claete leah* and speaks of 'the woodland clearing where burdock grows'. Note the earlier form also includes *inga* or 'the dwellers at', thus referring to the people rather than the place itself.

Coatham Mundeville

Two records of this name of note, as Cotum in the twelfth century and as Cotum Maundevill in 1344. This name features the Old English *cotum*, a plural of *cot* and referring to 'the cottages'. The addition here is manorial, a reference to the Amundevilla family who were here in the thirteenth century.

Beaumont Hill is undoubtedly the most common English place name derived from Old French, it describes 'the beautiful hill' with the later addition of Middle English *hyll*. Hall Garth describes 'the enclosure of or by the hall'.

The Foresters Arms is an early pub name, although today it refers to the Ancient Order of Foresters, a friendly society with courts in the UK and US.

Cocker Brook

Documented as Cocertun in the middle of the eleventh century, this British river name from *kukro* meaning 'crooked', a reference to its winding course.

Cockfield

A name describing itself as 'the open land frequented by the cocks (of wild birds)', derived from Old English *cocc feld*. This name is recorded as Kokefeld in 1223.

Esperley is a local name recorded as Esperdeslegh in 1230, a name describing 'the woodland clearing of a man called Aespheard'. This personal name is undoubtedly derived from 'aspen wood'. Understood to be a nickname it would not be used in any complimentary sense as aspen is amongst the softest of woods, making it virtually unusable.

Coniscliffe (High & Low)

Two places, with self-explanatory additions, having a common origin. This comes from Old English *cyning clif* and describes 'the king's holding at the cliff or bank'. The earliest record of this name dates from the middle of the eleventh century as Cingcesclife.

Carlbury describes 'the stronghold of a man called Carl'. Here the personal name comes from Old Scandinavian *carle* equivalent to Old English *churl* or peasant. Ulnaby Hall was built in the area previously recorded as Vluenebi in 1313, Ulneby in 1340, and Ulmeby in 1366. Here is another of Old Scandinavian origin, this speaking of 'the *by* or farmstead of a man called Ulfhethinn'.

Regulars at the Duke of Wellington will be aware this was the title taken by Arthur Wellesley (1769-1852). Best known for his military career, he later enjoyed political success as prime minister (1828-30) and foreign secretary (1834-5). Known as the Iron Duke, his name adorns more pub signs than anyone except Lord Nelson. The Baydale Beck Inn takes the name of the nearby stream.

Spotted Dog public house, High Coniscliffe

Consett

A name recorded as Covekesheued in 1183 and as Conekesheued in 1228. Here is a name combining Celtic (or possibly Pre-Celtic) *cunaco* and Old English *heaford* and telling us of 'the headland of a hill known as Conek'.

Standing alongside the River Derwent is a region historically known as Allensford. Today the name is seen in the road name of Allensford Bank and in the small area of parkland, however this have never been a well-populated area hence records are restricted to Aleynforth in the fourteenth century. While this is not overly useful in deciding upon the personal name represented by the first element, the suffix has remained unchanged until today. Here this is thought to represent a Saxon name and giving 'the ford of a man called Aella'. Cold Rowley adds an element indicating it was 'exposed' to a common place name describing 'the rough woodland clearing' from *ruh leah*.

Waskerley is recorded as Wascroppe and Wascroppe during the thirteenth century. There seems to be three elements here, a Saxon personal name Old Scandinavian *kjarr* and Old English *leah*, which together would refer to this as 'the woodland clearing in the marsh of a man called Wassa'. Greymare Hill takes its name from 'a grey boundary stone or mark', from Old English *grey mare*.

Mosswood is Old English but not recorded with the suffix *wudu* until the seventeenth century, every earlier record shows this comes from *mos ford* 'the mossy ford'. This can only mean the name was transferred to the woodland, thereafter becoming a place name. North of here is Shotley Bridge, this from Old English *Scotta leah* and a reminder of the influence from north of the border in 'the woodland clearing of the Scots'. Waskerley is probably describing 'the woodland clearing at the wet marsh'.

The Black Horse is a common pub name and an image which is clearly from an heraldic source. However the image is very popular in coats of arms and the problem is always to work out which of the many possible origins is correct, nearly always this proves impossible. The Cricketers is a common name for a village pub, most often showing the team were based at the pub, undoubtedly formed by the customers. While the coast is only ten miles away and the link to the sea is a possibility for a pub named the Anchor, the vast majority are a religious symbol.

The Freemasons Arms were originally stone masons, in modern times it is a friendly society who perform many acts of charitable work. Virtually every pub called the Grey Horse would be inspired by a favourite animal, one which would be born black and then develop an increasing number of white hairs as it matures. History shows the pub had a link with the horse for centuries, this is reflected in the names of many pubs including the Horse and Groom.

To prefix a name with 'Jolly' is an advertising ploy, it instantly suggests those frequenting the establishment are invariably happy and, in the case of the Jolly Drovers, specifically those who brought livestock to market. Britannia was the Roman name for Britain, although the name is a patriotic one referring to the image of the woman first seen on a medal or coin in 1665. The Black Diamond is a different reference to mining, not of diamonds but coal. The Manor House Inn occupies, as the name suggests, the former manor house. The Derwentside is a pub which is named from its position on the riverbank.

Cornforth

A name found as Corneford in 1196, this comes from Old English *corn ford* and describing 'the ford frequented by cranes or herons'. West Cornforth speaks for itself, although today the way both settlements have spread makes West Cornforth appear to the south.

Local names include Thrislington, from *ing tun* with an uncertain personal name and either 'the farmstead associated with a woman called Thrythhild' or 'the farmstead associated with a man called Thrythbeald'.

The Square and Compass are instruments used by a number of tradesmen, including the ancient skills employed by carpenters, joiners, and masons.

Cornsay

A name found as Corneshowe in 1183 and, similarly to the previous entry, comes from Old English *corn hoh* and describes 'the hill spur frequented by cranes or herons'.

Cotherstone

Domesday records this name as Codrestune in 1086, here is a name from Old English *tun* and a Saxon personal name and speaking of 'the farmstead of a man called Cuthhere'.

Coundon

Here is a name meaning 'the hill where cows are pastured' and coming from Old English *cu dun*. The name is recorded as Cundun in 1196.

Leasingthorne is a minor place name which has two equally plausible meanings. Either the first element is a personal name and thus 'the thorn tree of a man called Lesing' or this is *laes ing thorn* and 'the thorn tree associated with the pasture'. In the latter case this would act as a signpost.

The Sportsmans Inn is an advertisement of the most popular pastimes found inside the pub and activities arranged by the regulars. Durham Ox recalls the fabulous beast bred at Darlington which was almost two tons in weight at the age of six. Purchased for the princely sum of £250 but made a substantial profit when he took the animal on tour in a specially constructed vehicle. However it was this vehicle that also proved his undoing for it collapsed and injured the beast which then sadly had to be put down.

Cowpen Bewley

Listed as Cupum in the middle of the twelfth century, the basic name comes from Old English *cupe* in a plural form *cupum* describing 'the place of the coops'. These were wicker baskets used for the trapping of fish. The second part comes from Old French *beau lieu* or 'beautiful place'.

Minor place names here include the Batts, from *bat* and describing 'an island, low lying land subject to possible inundation'; Calfclose Bridge tells us it was 'where calf were raised'; Holm Fleet comes from *holm fleot* and speaks of 'the dry ground in a marsh by the tidal stream'; and the field name of High Fussack must speak of *foss* meaning 'ditch' rather than *fussock* speaking of a 'fat, troublesome person'. Salt Holme speaks of 'the dry land in the marsh with a salt works or where salt is sold'.

Coxhoe

Listed as Cockishow in 1277, here is a Saxon personal name and Old English *hoh* which describes this place as 'the hill spur of a man called Cocc'.

Tursdale has no early forms and, while the personal name is by no means certain, will probably represent 'the valley of Thor'.

A Scottish dialect term is seen in the name of the Kicking Cuddy Inn, the northern counties have also been influenced by this reference to a donkey.

Crook

Found as Cruketona in 1267 and as Crok in 1304, this comes from Old English *croc* and refers to 'the secluded corner of land'.

To the south is the hamlet of Bitchburn, a name where the second element is obvious but the first uncertain. The earliest record is from 1367 as Bicheburn, although the name is probably much older. If this is Old English *bec burna* then this represents 'the stream of the beech trees', with the first element influenced by the Middle English pronunciation. However the first element may represent a Saxon personal name and thus 'the stream of a man called Bicha or Bica'.

Oakenshaw is a common enough place name always from *acen scaege* and 'the enclosure of or by the oak trees'. Sunnybrow describes 'the top of the rise of a man called Sunn or Sunna'. Gin Mill is a slang term for a pub rediscovered in England following use in the United States. Rumby Hill is recorded as Ronundby in 1382, this describing 'the *by* or farmstead of a man called Hromundr'.

The Green Tree is a pub name which takes the simple imagery to reflect a prominent local tree. Advertising fishing and also the menu, the Brown Trout is a simple pub name, while the Burn Inn refers to the stream. The Crown is a simple and easily recognised

Crook still proudly proclaims itself a market town

Travellers Rest public house, Crook

Wonderfully imaginative sign at the Colliery public house in Crook

symbol which, even to the illiterate, would signify the monarchy and a patriot. The White Swan's origins depend upon its age, recent examples will refer to the majestic bird, further back it is an heraldic connection, chosen by many families and nobles, including the Vintners' Company, Poulters' Company, Musicians' Company, earls of Essex, and Edward III.

The Brewers Droop is an unusual if amusing pub name. The term is a colloquialism for erectile dysfunction through alcohol intake. As names and signs are adverts to prospective customers, this seems an odd promise for the landlord to make. Ye Olde Horseshoe tells us it was named for the association with the horse or blacksmith, was represented by a symbol which is instantly recognised and could never be seen as anything else. We also know, despite the name, it cannot be more than a century old for 'Ye Olde' is an American idea which came to England in the post-Victorian years of the early twentieth century.

The Golden Fleece is associated with the story of Jason and the Argonauts of Greek mythology, however as a pub name it symbolises the Knights of the Golden Fleece, an order of chivalry founded in 1429. Reginald Farrer was a noted early twentieth century botanist and writer and is remembered by the Farrers Arms. The Royal George Inn is not named directly from a monarch but after a ship.

Croxdale

Croxdale is a minor name recorded as Crokesteil in 1335 and Croxdaill in 1570. Here the latter example, and of course the modern form, suggest the suffix *dael* or 'valley'. However the fourteenth century form points to a Middle Scots dialect word *tail* which describes 'the irregular boundary of land jutting out from a larger area', here with the personal name Krokr. Sometimes we can deduce further information by defining a place name. While the personal name is clearly Old Scandinavian, this suffix is almost certainly evidence of it being worked by someone from north of the border.

In the coaching era from the seventeenth century it was as important for travellers to know where the coaches dropped and picked up as it is for bus passengers to know where the bus stop is today. This is the sole reason for the naming of the Coach and Horses public house. Although today there is no traditional sign outside the establishment, without such the name no longer works. The Honest Lawyer is an old joke, usually shown as a headless man wearing silks, thus saying the only honest lawyer was a dead one.

D

Dalton-le-Dale

Found in a document dating from the eighth century as Daltun, this name comes from Old English *dael tun* and refers to 'the farmstead in a valley'. The suffix is clearly influenced by a French scribe, while the ending has the same meaning as the first element. Dawdon shares the same first element, this being the hill by the valley'. Murton is from *mor tun* 'the farmstead by the marsh'.

Darlington

Listed as Dearthingtun in 1009, this name features a Saxon personal name and Old English *ing tun* and speaks of 'the farmstead associated with a man called Deornoth'.

Here we find Blackwell to the south, a name from *blaec wella* and describing 'the dark spring or stream'. Cockerton stands on a small brook called the Cocker, hence this is 'the farmstead on the Cocker'. Bondgate is an historic street showing it occupied land owned by the bondsmen of the town. Harrogate is a district to the north of the town, a name referring to 'the road with a heathen temple' from *hearg geat*. Oxenhall is easy to see as 'the nook of land where oxen are kept'. Ulnaby is found to the northwest, a name of Old Scandinavian origin where *by* follows a personal name and tells of 'the farmstead of a man called Ulfwine'.

Whessoe is found as Quehowe in 1333, as Whessowe in 1345, and Quesshowe, Qwhessow, and Wessow in the fifteenth century. This comes from Old English *hoh* and a Saxon personal name giving 'the hill spur of a man called Cwen'. Baydales is found in 1190 as Badele, in 1340 as Bathelspitel, and in 1784 as Bedelbeck. The first element is

Darlington sign

certainly a personal name, probably a pet form of a name and thus given as 'the dale of a man called Basa'. These later records refer to the same place but to different features, the hospital and the stream respectively. Polam is derived from *pol ham* Old English for 'the homestead by a pool'.

Many pubs were named after prominent trees, a good idea in England's green and pleasant land when a tree would be an excellent marker. This was continued in later years and the Copper Beech is a good example, this tree's very dark leaves make it quite obvious even when surrounded by other species. In the eighteenth century what is now the Quakerhouse was a yard where coopers and blacksmiths worked. While there seems to be no direct link to the Quaker movement, the name comes from the previous occupants who called their establishment the Quaker Coffee House.

The Tawny Owl is a simple and easily recognised image, and more likely to have been the reason for the name of this pub rather than any symbolic reason. The Nags Head dates back to the time when the horse was available for hire, transporting the traveller to the next pub in the network. Albion is chosen for a pub name for it shows patriotism, this being the Latin name for England.

Weaving dates from prehistory, however the Shuttle and Loom recalls the importance of flax-weaving to the town. Another trade is seen in the Boot and Shoe, this establishment off the marketplace would suggest repair rather than manufacture on a small scale. The Golden Cock is heraldic, used by numerous tradesmen, including the Courage brewery which has virtually ended its use as a pub name and makes tracing the origins of a specific example quite impossible.

With the sign outside the Glittering Star showing a large star hanging over the local church, the meaning is clearly religious and an invite to the nearby place of worship. The Hope Inn is also linked to Christianity. Lying close to the border with Scotland, the influence from north of the border is seen in the Burns Tavern, named after the Bard of Scotland, Robert Burns.

The Railway Tavern tells us it was not only found near the new mode of transport but could not have existed prior to its arrival. The Caledonian is also a reminder of the railway, the Caledonian Railway later became a part of the London Midland and Scottish Railway, however the sign here shows a Great Western Railway King class locomotive! The Falchion is named after a curved broad sword, said to either have been that used to kill a mythical beast who preyed on

Probably Darlington's most famous sight, the clock tower at the railway station

the locals or the weapon which executed Peasants Revolt leader Wat Tyler in 1381. The Speedwell recalls *HMS Speedwell*, one of more than a dozen vessels so named, the example shown on the pub sign is in full sail. The Dalesman refers to those to the south, one from the Yorkshire Dales. The Hole in the Wall is a common name, one with nearly as many origins as there are examples. Clearly it refers to a narrow opening, yet this could be anything from a peep hole or serving hatch, to a narrow corridor or even an alleyway.

Outside the Pennyweight hangs a delightful sign. Here a large model of a balance scale hangs which, together with the name, tells us this place was associated with those who used this system of weights – apothecaries and jewellers. A pennyweight was literally the weight of the early penny, 1/240th of the weight of a troy pound, with 240 pre-decimalisation pennies to the pound sterling.

The Havelock Arms takes the name of General Sir Henry Havelock, hero of the Indian Mutiny and the Siege of Lucknow, who died two months later. A prize shorthorn bull gave his name to the Comet after he fetched the remarkable price of one thousand guineas. A name such as the Plastered Parrot can be placed in two categories, both humour and aliteration, the latter the advertiser's dream.

Britannia was the Roman name for Britain. The female figure and her shield now associated with the name appeared on Roman coins but was not seen on a truly British minting until 1665 and that was a medal. The present figure was modelled on Frances Stewart, who came to prominence as the mistress of Charles II and was later titled Duchess of Richmond. From the trade rather than a surname comes the Slaters Arms. Another trade was the inspiration for the Tap and Spile, an indirect reference to the ale kept in casks by the inn-keeper. The tap is self-explanatory, the spile is the wooden peg which is removed to allow air to enter the cask and replace the volume of ale drawn off through the tap, then replaced to restrict the release of carbon dioxide from the ale. The Tanners Hall is a reminder tanning took place here.

Pubs called the Green Dragon are often depicted with St George battling a dragon. These are signwriters' interpretations and have nothing to do with the true origin. Here the sign shows the dragon in its true light, as an heraldic image and representative of the earls of Pembroke, major landowners of England. Darlington's link to the railways is seen in Ye Olde Wheel Tapper Inn, the job of the wheel tapper was a lowly but important task checking for cracks in the metal wheels. Note the pub cannot be as old as the name would suggest for Ye Olde was not seen until the late nineteenth century. It has been imported from across the Atlantic and makes no sense for 'Ye' is not 'the' but 'you'.

Denton

The earliest surviving record of this name dates from around 1200 and is exactly the same as the modern form. This is from Old English *denu tun*, a common place name referring to 'the farmstead in the valley'.

Derwent (River)

A Celtic river name with at least four examples in England. All of these are derived from the description of 'the river where oak trees grow abundantly'.

At over forty miles long the length of its banks are clearly twice this. As the nature of the river and the landscape changes considerably over its course we would never expect to find the same fauna growing there, thus it is logical to assume the river cannot have been known by this name for its entire length – remember travelling was not commonplace until comparatively recently. Hence the oak trees mentioned in the river name would only have grown at certain points and it must have been at one of these points that the name was first recorded as such. Indeed the name for the entire river almost certainly came from an early map.

Dipton

Recorded as Depeden in 1339, this name comes from Old English *deop denu* and describes 'the deep valley'.

The Prince of Wales is a royal name which most often refers to the future Edward VII who had held this title for longer than anyone until Prince Charles. The Flint Hill Inn advertises its location by adopting the local place name.

Don, River

For a Celtic name this is quite common, there are other examples in Yorkshire and another flows into the North Sea off the east coast of

Scotland. Common place names nearly always point to a very simple meaning and none can be more simple for a river name than 'water'.

Obviously water is a fundamental requirement of all life on the planet and must have been one of the first words ever coined by mankind. Nearly every language in Europe and the Indian sub-continent have a connection, this Indo-European language group having a theoretical Proto-Indo-European tongue which, while it will never be known, can be shown to be related through some ancient languages.

For example with the River Don or 'water', we are aware of the Celtic or British *dana* meaning 'water', itself related to the 'water' seen in the name of one of Europe's and the world's great rivers the Danube. Further east the great ancient language is Sanskrit. Here we find *danu* which meant 'rain, moisture'. It is not difficult to see how all these would be derived from a mother tongue.

Durham

The county town, and thus the county, has a name derived from Old English *dun* and Old Scandinavian *holmr*, which together describe 'the island with a hill'.

Before looking at the present street names of the city, it is worthwhile examining those names no longer in use for they reveal something of Durham's history. The streets which brought travellers to the city's gates were known as Walkergate, Smithgate and Fleshergate. Here the suffix *geat* does not refer to the barrier across the route but the way itself. The names described 'the way of the cloth workers', the way of the metalworkers', and 'the way of the fleshers or butchers' respectively.

Other trades were seen in Souter Peth, 'the street of the shoemakers'. Potters Bank was where medieval potters produced their pottery. Pit Lane and Old Pit Terrace provided accommodation for miners at the nearby colliery. Diamond Terrace was another reference to coal, also known as 'black diamond'.

The skyline of Durham is dominated by its cathedral, hence it is no surprise to find religion looms large in the names of the streets. Mount Joy Crescent is a delightful name indicating the feelings of those reaching the terminus of their pilgrimage, while Palmer's Garth marked the enclosure set aside for the pilgrims. Hallgarth Street takes the name of the farm belonging to the cathedral, Anchorage Terrace marks the home of a hermit or anchorite hence the name.

The medieval hospital associated with St Giles Church provided a name for Gilesgate. With St Giles being the patron saint of lepers this may well indicate this was a leper hospital. Margery Lane takes the pet form of the saint associated with St Margaret's Church. Quarryheads Lane may seem more industrious than religious however the link refers to the sandstone quarried in order to build the cathedral.

Street names tend to stick long after the reason for the name has disappeared. Stones once found around what is now known as Baxter gave the place a name. This is a corruption of bakestone, those used to line ovens and which also gave a name to Quarry House Lane. Similarly a medieval oven house existed on the now corrupted name of Owengate. Bakehouse Lane was another named for its bakeries. Toll House Lane marks the site of the toll levied on those travelling to Durham during the coaching era, perhaps they were heading for Whitesmocks where an inn took its name from the long coats of the waggoners.

A well dedicated to St Mary is the most likely origin for Mayorswell Street, although there is a record of a family called Maire who held land here in the fifteenth century. As a source of water was very important it is no surprise to find other

County Durham styles itself 'the Land of the Bishops'

Pity Me sign

The Lambton Hounds Hotel at Pity Me

references in Springwell Avenue, Springwell Road, Springwell Hall, and Framwellgate. Because the development was planned from a grid with no named streets, east-west were letters and north-south numbers. Eventually these were replaced by alternative names, with the exception of Ninth Street, which still appears on the road sign to this day.

Minor names include Butterby, listed as Beautroue in 1345 and as Buttry in 1355, the latter form (and the modern) would normally be defined as from Old English *buttere* and Old Scandinavian *by* and describe 'the farmstead where butter is obtained'. However the earlier record shows this is probably a corruption of Old French *beau trou* 'the beautiful spot'. Of course the reverse could also be the case.

Finchale is recorded as Wincenhale, Winchanheale, Pinchala, Finkale, Finchale, Feynkhale, Fynghall, Fynkhalgh, Fenkal, and Fencall up to the fifteenth century. There is no doubt this is Old English *finc halh* or 'the nook of land frequented by finches'. Houghall is from *hoh halh* telling of 'the nook of land on or by a hill spur'. Records of Kepier are extremely varied, making definition difficult but most likely a reference to 'the weir of a man called Cyp or Cyb'. Frankland appears as Frankleyn in the fifteenth century, clearly a surname.

Baxterwood is recorded as Bakestaneford in 1199, Bacstanford in 1300, and as Bxstanford in 1472, all of which point to an origin of 'the ford where backstones are removed'. Burn Hall was built near the *burna* or 'stream'. Redhills is from *red hyll* 'the hill with red coloured soil'. Early records of Relley may show this would be expected to have normally become Ridley, this being 'the woodland clearing cleared of brushwood'. Risebridge is first recorded in the fourteenth century, this describing 'the bridge of a man called Hrisa' or 'the bridge by the brushwood'. Causey Park takes its name from the ancient paved causeway which forms its eastern boundary. Harbour House comes from Middle English *harbarwes* 'the shelter or lodging'. Whitwell is not difficult to see as 'the white or bubbly spring'.

Pubs in the city include the Durham Light Infantryman, clearly named for a member of the regiment of the British Army founded in 1881 and disbanded in 1968. Always a good idea to refer to the product and those that produce it, hence the popularity of the Brewers Arms. Another reference to the product is seen in the Hogshead, the pub bears the name of the large cask for wines and beers of no set capacity, while the Pot and Glass refers to the containers of the drink. The Woodman Inn began as a reminder of the importance of this tradesman to all settlements, irrespective of the location. Alongside the river at Elvet Bridge we find the Swan and Three Cygnets, for once a 'swan' pub name where we can be certain it is not heraldic but refers to the waterfowl. The New Inn may seem simple enough, however it should be realised this also tells us there was an earlier pub in the vicinity and should be seen as the 'newer inn'.

Themes are not restricted to the modern era as this early Durham development shows

The Shakespeare Tavern is named after the Bard of Stratford-upon-Avon, the most famous of all wordsmiths has more pubs named after him than anyone except for Lord Nelson. That Durham University can claim to be the oldest outside Oxford and Cambridge means a pub called the Varsity was almost inevitable. Ye Olde Elm Tree tells us it was named from a landmark tree and that it is not that old for 'Ye Olde' is from

Colpitts Hotel in Durham

the US and arrived here around the end of the Victorian era.

It is no surprise to find the Stonebridge Inn alongside the river. The Angel Inn reminds us there was a long-standing connection between the local pub and the church. This seems odd today but when almost everyone worked the land the only time they met with anyone other than their family or colleagues was at their place of worship or refreshment.

The Fighting Cocks recalls the specially bred birds unleashed upon one another in the pit, spectators betting on the outcome. Brought to England by the Romans it was banned by Cromwell in 1653 which only succeeded in sending it underground until sterner measures were taken to eradicate it during the nineteenth century. The name of the Coach and Eight is a clear reminder this was a former coaching inn. Salutation is an early religious term, a reference to the Annunciation when the Archangel Gabriel visited the Virgin Mary to announce she had been chosen to give birth to the infant Christ. The Rams Head is representative of the wool trade on which England's wealth and power were based.

The Marquis of Granby is a name found across England. It remembers John Manners (1721-70), Colonel of the Royal Regiment of Horse Guards and commander-in-chief of the British army. As the conflicts ended and his men returned home, he set many of them up as inn-keepers and, to mark his generosity, they in turn named their premises after their former commander. The Jovial Monk shows the close location of the church, while suggesting a good time could be had within. Other welcomes are offered to passers-by of both the Wayfarers Inn and the Ramblers Rest. The Davy Lamp extends a welcome to miners. The Hare and Greyhound represents hare coursing, a sport which was more about the chase than capture but was banned with other blood sports in 2005. The Greyhound's name is quite different, this is heraldic and points to the dukes of Newcastle, major landowners in the area.

Particularly in rural areas poachers provided a reliable source of fresh meat. Such would often be offered for sale or barter at the local pub and, in order to hide their capture, they had large pockets sewn inside their jacket. The name of the Poachers Pocket was inevitable as a pub name. Early examples of the Wishing Well would point to a nearby spring, however in later names the 'well' is more likely to be a keg of ale. The Village Inn would have been, with the church, the focus of village life and referred to as such so often the name would inevitably have stuck. Next to the village pub was inevitably found the village blacksmith, many pub names reflect this and the Shoes is simply an abbreviation of horseshoe, an advertisement to re-shoe the travellers' mount.

E

Easington

A name recorded as Esingtun around 1050, this comes from a Saxon personal name and Old English *ing tun*. This name speaks of 'the farmstead associated with a man called Esa'.

Flemingfield takes the name of John le Fleming, a Flemish man who was granted this land before the fourteenth century. Pespool is listed as Pesepole in 1316, the name describing 'the pool near where peas are grown'. Edderacres is found as Etheredesacres in 1314, this being 'the agricultural land of a man called Aethelred'. Thorpe is a common place name, this describes 'the outlying farmstead' and would originally have been used as temporary accommodation during the growing season when tending to livestock or crops.

The Half Moon Inn utilises a very simple and easily recognised image, inviting patrons during hours of darkness, a reminder of the times when licencing laws reduced opening hours and still the busiest time, although it began as an heraldic symbol.

East Rainton

First seen around 1170 as Reiningtone, this is Old English *ing tun* and a Saxon personal name and speaks of 'the farmstead associated with a man called Raegen or Regna'. The addition is to distinguish this place from West Rainton.

The minor name of Moorsley is found as Morlawe and Moreslawe around the end of the twelfth century, this shows the suffix is not *leah* but *hlaw* and thus 'the burial mound of a man called Mor'.

Ebchester

A name which tells us it was 'the Roman stronghold of a man called Ebba or Ebbe'. The name features a Saxon personal name and Old English *ceaster* and is recorded as Ebbecestr in a document dated 1230.

The Derwent Walk Inn stands near the river, the name defined under its own entry.

Edmondsley

Found as Edmansley, Edmond-esley, Edmansle, and Edmanesley in the earliest forms, this features a Saxon personal name and Old English *leah* referring to 'the open land of a man called Eadmund'.

Locals enjoy a drink in the Charlaw Inn. Originally known as the Blackhouse Inn, the present name was taken from the nearby fell.

An unusual sign at Edmondsley

Edmundbyers

Found south of the Derwent Reservoir, this comes from a Saxon personal name and Old English *byre* and speaking of 'the cattle shelter of a man called Aedmund'.

Locally we find Roughside, a name which still virtually describes itself as 'the scrubland on the hillside'. A combination of Middle English and Old English *petemos ake* gave us Pedam's Oak or 'the oak tree by the peat moss bog'. Swinburn has changed little since the thirteenth century record of Swineburn, this being 'the stream frequented by wild boar'.

Charlaw Inn at Blackhouse, near Holmside

Egglescliffe

Found as Ecclesclive in 1197 and as Egglesclive in 1340, this is Old English *ecg cliffe* and describes the '(place at) the edge of the cliff".

Local names include Aislaby, recorded as Aslachesbi in Domesday it is of Old Scandinavian origin and describes 'the farmstead of a man called Aslac'. Urlay Nook is thought to be derived from 'the nook of land of Eorl's meadow'. Trafford Hill features a common name, one from *treow ford* and telling of 'the ford marked by a tree' and which also tells us this must have been a very conspicuous tree. Newsham has changed a little since its early days of *niwe hus* 'the newer house or houses'.

Trefford appears as Treiford in 1189 and as Trafford in 1649. This is difficult to define, for early forms are few, but probably comes from Old English *treow ford* or 'the ford marked by trees' or perhaps this should be understood as 'the ford made from timber'.

Field names include Fogg Close, *fogge* literally describing 'the aftermath' and is understood to refer to 'the vegetation left in the field after the harvest', which would have provided food for livestock. Caster Close features *caster*, a job where one 'casts coals from ships into keels'. Clearly this has nothing to do with the field and must refer to ownership, probably a small holding worked by his family.

Eggleston

In 1196 this name is recorded as Egleston, this name comes from Old English *tun* and a Saxon personal name and tells us it was 'the farmstead of a man called Ecgel'.

When looking a pub names a 'tun' is something altogether different. Although it refers to a large barrel containing around 250 gallons, the name of the Three Tuns is actually heraldic, appearing in the arms of the Worshipful Company of Vintners and Worshipful Company of Brewers.

Elton

A name of Old English derivation, where the Saxon personal name is suffixed by the common element *tun* and refers to 'the farmstead of a man called Ella'.

Field names of Elton include the delightfully-named Old Prankey, *prank* is an old dialect term meaning 'a pleat or fold (of land)'; Mustard Garth is a place 'where mustard grows'; and Whinney Ford is at a place 'overgrown with gorse'.

Elwick

Recorded as Elevet in 1199, as Ellewi in 1211, as Elwyk in 1291, and as Elleswik in 1314. This represents a Saxon personal name and Old

English *wic* and refers to 'the specialised farm of a man called Ella or Elli'. Almost without exception the speciality would be dairy produce.

Locally we find the name of Amerston, the suffix is sufficient to show this is of Old English origin and describes 'the farmstead of a man called Amund, or Aemund'. Stotfold is a name found elsewhere, all from Old English *stod fald* or 'the stud farm for horses'. Benknowl describes itself as 'the hill where beans are grown'. Worset Lane is from *weard setl* referring to 'the look-out place'.

The Tofts is an Old Scandinavian word meaning 'buildings, plots'. Three names, High Burn Toft, Middle Burn Toft, and Low Burn Toft, add *brunnr* to *toft* and thus describe 'the curtilage by a spring'. Pudding Poke links words pointing to 'a sticky or muddy place' and 'cul-de-sac' respectively, we must assume it fitted both descriptions.

Field names include Jaw Blades, named for its shape, as indeed is Three Nooked Farm. Loaning House takes its name from *loning* a Middle English word meaning 'the lane'. Stotfold is from *stod feld* 'the open land where horses are reared'.

Escomb
Recorded as Ediscum in the tenth century, this is derived from a plural form of Old English *edisc* and describes the '(place at) the enclosed pastures'.

The Saxon Inn is an ideal name for a pub in a book where many of the names remember that era of English history. It is also a reminder of the local church, one of the oldest churches in England. Built between 670 and 675, it was constructed from stone 'recycled' from the old Roman fort at Binchester. Set in a distinctive circular churchyard, this place of worship has been in almost constant use since it was first built.

Esh
From Old English *aesc* and speaking of the '(place at) the ash trees'. This name appears in the twelfth century as Esse.

Flass is recorded as Flash and Flaskes, from Old French *flasque* 'a pool' and used in Old English as *flassks* and meaning 'low lying boggy land'. This usage is mostly restricted to the northeast of the land, more common is the dialect *plash* with the same origins and meaning. The Flass Inn copied this name, while Esh Winning gave its name to the Winnings public house, despite the clear link to horseracing.

Malton has two possible origins, either a personal name and 'the farmstead of a man called Maela' or 'the farmstead with a marker' from Old English *mael*, although what form that marker took and what it was marking will undoubtedly never be known. Biggin is found as Biging in 1490, this coming from *bygging* and referring to what must have been a very substantial and impressive 'building'. Heugh comes from Old English *hoh* or 'spur of land'. Ushaw has confused Old English *sceaga* and Old Scandinavian *skogr*, the latter being the true origin while the former is clearly seen as part of the name. This is simply 'the wood of a man called Ulf'.

Evenwood

A name found in the middle of the eleventh century as Efenwuda. This name comes from Old English *efen wudu* and describes 'the level woodland'.

Morley is a local name derived from *mor leah* or 'the woodland clearing at or by the moorland'. Ramshaw is seen as Ramsale in the earliest surviving records, a name from Old English *hraef scaga* and telling us it was 'the woodland copse frequented by ravens'. Recorded as Ramshale in 1382, Ramshaw is probably *hrafn haugr* and 'the mound frequented by ravens' although the first element is also used as a personal name.

The Bridge Inn could not be more aptly named, for it stands facing the bridge crossing the River Gaunless. The Travellers Rest reminds us why the country inn developed, as a place of refreshment.

F

Fatfield

Fatfield is fairly, although mostly as a common minor name, normally found labelling a field, it describes a fertile area literally 'the open land producing good profit'.

Minor names include Picktree, this would have spoken of 'the tree of a man called Pice' and probably a boundary marker.

Fendrith

A name also found in that of Fendrith Hill, that this has been transferred to the hill from the wetland to the south is clear from its origins. Here the 'd' is intrusive, simply born of pronunciation, and plays no part in the etymology. This is Old English *fenn rith* and describes 'the bog with a stream'.

Ferryhill

The earliest record dates from the tenth century as Feregenne and as Ferye on the Hill in 1316. Old English *fergen* refers to 'the wooded hill' with the later addition of *hyll* which also means 'hill'.

Nearby Garmondsway is named after the road known as 'the way of a man called Garmund'. This was the road taken to the shrine of St Cuthbert, indeed there is a written record of King Cnut (or Canute) walking along here in bare feet on his pilgrimage to pay homage at the shrine. Raceby has not changed since the earliest known record which dates from 1345 and describes 'the *by* or farmstead of a man called Hreithr'.

Windleston's early forms swing between the suffix *dun* 'hill' and *tun* 'farmstead', the modern name misunderstands the name completely and gives it as *stan* or 'stone'. The remainder is a Saxon personal name and, with the topography favouring *dun*, hence this gives 'the hill of a man called Waendle'. The Breaks refer to the 'brakes or thickets of bushes, brushwood and briars'. Thrundle Farm is a name seen since the late fourteenth century and describes 'the dale of a man called Thurwine'.

Mainsforth is found as Manceforth, Maynesforth, Mannesforthe, and Manesforth, this name may well come from Old English *maegen ford* 'the strong ford' in which case it will refer to the current. However there is a possible possessive 's' here and therefore could well be 'the ford of a man called Maegen'. While it is impossible to tell from the forms available, it should be realised the personal name would always be preferred by toponymists for fords invariably have a personal name for a first element. This is understandable for fords are not entirely natural features but have almost been created by clearing of vegetation and embedding stones in the bed of the stream to provide a firmer and shallower surface.

Public houses called the White Horse increased in popularity in the eighteenth century. This coincided with the coming accession of the House of Hanover, the image representative of the family. The Surtees Arms recalls 1870 when Robert Surtees purchased much of the meadow and pasture around the manor house. An heraldic image, the Greyhound recalls the dukes of Newcastle. The Postboy Inn remembers those carried mail and used inns as collection and drop-off points. Probably suggested by a former military man, the Flintlock Inn recalls the standard military rifle used from the seventeenth century until the nineteenth when the cartridge was introduced.

Fishburne

This name is documented as Fisseburne around 1190, this name comes from Old English *fisc burna* and describes 'the stream where fish are caught'.

Locally we find Blakston, a name found as Blekeston in 1204, with other records of Blaichestun, Blaykeston, and Blaxton. This most likely represents 'the farmstead of a man called Blaec'.

Foxton

A name which is derived from Old English *fox denu* and describes 'the valley frequented by foxes'. This name is found as Foxdene around 1170.

Framwellgate Moor

Listed as Framwelgat in 1352, this comes from Old English *fram wella* and Old Scandinavian *gata* and describes this as 'the street by the strongly gushing spring'.

Here the pub is the Happy Wanderer, the name refers to a group of mine workers who came from Yorkshire to Durham in the 1930s following closure of the Yorkshire pits. The sign features a cheerful miner, while the pub is home to an excellent collection of miners' lamps. This may be an indication that one of the miners came here to work the mines and ended up being the landlord.

Frosterley

A name found in a document from 1239 as Forsterlegh, which comes from Middle English *forester* and Old English *leah* which describes 'the woodland clearing of the forester'.

Minor names here include Bollihope, a name from a Saxon personal name and Old English *hop* speaking of 'the valley of a man called Bol'. Pye Close features the dialect word *pie* for a 'magpie', probably suggesting there were many of these birds here. Riddyng House takes its name from Old English *hryding* describing 'a clearing'.

G

Gainford

Recorded as Gegenforda around 1040, this name comes from Old English *gegn ford* and describes 'the ford on a direct route'.

Minor names include Heddlam, which has changed little since the earliest record of Hedlam and probably represents 'the homestead of a man called Aethel'. Selaby is recorded as Selbye and Selysby, a combination of Old English *sealh* and Old Scandinavian *by* and describing 'the farmstead where sallow grows'. Piercebridge describes 'the bridge of a man called Piers', not 'the bridge built on piers' nor 'the bridge of the priests' as suggest by the sixteenth century record of Preistbrigg.

The evolution of the name of Alwent is a little complex. It must have been named from the small stream which flows here, today known as the River Alwent. This was not the original name of the stream but comes from the settlement, a process known as back formation. However the settlement has lost a part of its name, this would have started out as Alwenton 'the *tun* or farmstead on the River Alwen or Alwyn', much as we find Alwinton on the River Alwin. This would represent a Celtic river name, related to Gaelic *aluin* and Welsh *alwyn* this represents 'the bright or clear stream'.

Selaby is recorded as Selebi in 1197 and as Seletby in 1322. While there is no doubt the name is Old Scandinavian, with the common suffix *by*, the first element is less certain but is almost certainly a personal name. The problem here is we have no knowledge of any name which fits with the known examples. This is not particularly unusual and is a good clue to this being a nickname. Hence we look for a Scandinavian phrase which would fit with the known forms, while also making some sense. Here the most plausible origin is found in the Saxon poem the Lay of Maldon in which the Vikings are referred to as *sae-lida*, the nickname being *Sae-lithi* and coming from *haf-lith* to give 'the farmstead of the sea traveller'.

Summerhouse comes as no surprise referring to 'the house used in summer', however it is rather odd to find a small-ish building creating a place name, albeit a minor one. Headlam is recorded since at least the twelfth century, this being 'the homestead of a man called Heddel'. Heighley Hall was built on land already named as 'the high woodland clearing' from Old English *heah leah*.

Marwood is recorded as Marawuda in the middle of the eleventh century and as Marwode in 1335. This probably represents Old English *mara wudu* 'the

Summerhouse welcomes us

larger wood'. No man in the history of England has had more pubs named after him than Lord Nelson, arguably our greatest hero.

Gateshead

Seen as Gatesheued in a docuemnt from 1196, this is derived from Old English *gat heafod* and describes 'the headland of the goats'.

While the place names show an unwritten history, street names can reveal themselves as pointers to more recent times when our towns and cities grew rapidly. For example the Redheugh Estate was owned by the Askew family from 1748 to the late nineteenth century and who gave their name to Askew Road. Bewick Road remembers the wood engraver Thomas Bewick, who lived and worked in Gateshead all his life (1753-1828). Lord of the manor of Gateshead for ten years from 1716 was William Cotesworth, who may not appreciate having the spelling of his name appear wrongly on the signs at the ends of Coatsworth Road.

From 1730 to 1857 the Ellison family were lords of the manor, hence the name of Ellison Street. As MP for Newcastle from 1812-30, Cuthbert Ellison is less well known for his generosity in establishing and contributing to local charities. Built in 1851, Ellison Square has identical origins. Deckham Street is a part of the estate owned by Sir Thomas Deckham in the early seventeenth century. Formerly known as Collier Chare, Jackson's Field and Jackson's Chare are first seen in the early eighteenth century. Both these refer to Henry Jackson, steward on the estate of the Gerard family, whose name is seen on the signs at the end of Jackson Street. Joicey Road takes the surname of the family resident at Whinney House from the middle of the nineteenth century.

Pipewellgate shows how this was where wooden pipes were used to bring water into the town, hence this refers to 'the way of the piped stream or spring'. The similarly named Oakwellgate describes 'the way of the spring sheltered by an oak tree'. Saltwell warns this water was undrinkable, however it could be useful as 'the salt water spring' would have provided priceless salt for preserving meat and the making of cheese.

Shipcote is a corruption of Sheepcote, easily seen as 'the cottages of the shepherds'. Sheriff Hill is found by the thirteenth century, a reference to the Sheriffs of Newcastle who would meet the judges from Durham at the assizes here. The name has since been transferred to Sheriff's Highway, itself previously known as Sodhouse Bank which referred to the turf huts built by tinkers. Wrekenton was claimed to have been named by antiquarian John Hodgson, who wrote his inspiration was the Wrekendyke Roman road. There is nothing to prove this was the case, indeed we would normally expect the place to give a name to the road.

Locally we find Dunston, a hamlet with an Old English place name meaning 'the farmstead of a man called Dun'. The earliest record of Eighton Banks is as Ayketon around the thirteenth century, this shows the name comes from Old English *ac tun* 'the farmstead by the oak trees'. Felling is recorded as de l' Felling and Felligwater, both found in 1351, and quite simply informing us this was 'where trees were felled'. Follonsby is clearly an Old Scandinavian suffix *by* following a Saxon personal name and thus 'the farmstead of a man called Fullan'.

Hedley is a simple enough name, from *haeth leah* this describes 'the woodland clearing by the heathland'. Heworth, recorded since the twelfth century, is from *heage worth* and describes 'the hedged enclosure'. Lamesley is first recorded in 1291, where it appears exactly as it does today. The place name probably shows the personal name Leman, a corruption of Leofman and giving 'the woodland clearing of a man called Leofman'. Marley, seen as Merlei and Marlei in the thirteenth century, this is from Old English *maer leah* and describes 'the woodland clearing at a boundary'. Pelaw is a district with a Saxon personal name and *hlaw* 'the mound of a man called Paelli'.

Ravensworth appears in the fourteenth century as Raffen, Raffenswoth, and Raffensholm which point to a modern origin of 'the enclosure of a man called Hraefen' or possibly the first element is *hraefen* 'ravens'. The alternative suffix of *holmr* tells us this was also known for its 'dry land in marsh'. Swalwell stands at the confluence of the rivers Tyne and Derwent, a permanent wetland which would be sure to attract insects and the birds that feed on them, hence this is seen as 'the rivers where swallows are seen'.

Wardley comes from Old English *weard leah*, 'the woodland clearing of the watchman or watch point'. Redheugh, recorded as Redhoghe in the late twelfth century, is from *red hoh* 'the spur of land where reeds grow'. Darncrook is an Old English name describing 'the hidden or remote crook of land'. Pountey's Bridge, which no longer exists, took its name from Middle English *pount* and also means 'bridge'. Friar's Goose is not what it seems at all, for it has nothing to do with either religion nor anseriformes, this is *Eryngium campestre* – a plant commonly known as Field eryngo which we must assume was routinely gathered here. A thorny perennial with tough, whitish-green leaves which are attacked by the gall fly. It was and is used in herbalism for the treatment of coughs, whooping cough and urinary infections. Formerly the roots were candied as sweets and roasted as a vegetable.

Colton comes from *col tun* or 'the farmstead of the charcoal burners'. Prior to the extraction of coal from the ground, 'coal' was produced by a slow burn of wood to remove all traces of moisture and other impurities leaving very little but carbon. To achieve this the logs were cut to a regular size, stacked in layers to form a rough mound which was then covered with turf leaving only a small hole for the smoke to escape. It took experience and skill to achieve the fine balance where the wood smouldered but not consumed by flame, nor would the slow burn be extinguished due to lack of oxygen. Thus the charcoal mound would need to be watched, both night and day, until the process was completed. Staying awake for 72 hours while staring at a virtually inanimate object was guaranteed to send anyone to sleep. Hence the charcoal burners took to sitting on a stool with a single leg, thus when sleep did come they would fall and quickly awaken. We still refer to going to sleep as 'to drop off''.

Gaunless, River

Listed as Gauhenles in the twelfth century, as Gawenles in 1242, and as Gaunles in 1291 there is some dispute as to the origins of this name. Generally thought to be a Celtic or British river name, we know of no related word in any of those modern languages closely related, such as Welsh, Cornish and Breton, which are the keys to understanding the early tongue of our islands. Normally Celtic river names are simplistic in the extreme, most speak of 'water' or 'river'. This tends to cast doubts on the alternative Middle English *gaghenles* meaning 'useless', especially as there is no known Old English word bridging the two languages. However this presumes the English words are related to the earlier Celtic, which is only sometimes true. Furthermore just because we have no knowledge of an Old English equivalent word does not conclusively prove none existed, nor does it show the Middle English *gaghenles* is not the origin. If indeed this is 'the

River Gaunless

useless' river then we can only think this must refer to it being unreliable as source for fish and/or plants.

Greatham

Early records include Grytham, Gretham, Grethame, and Greteham. This is from Old English *greot ham* and describes 'the homestead on the gravelly land'.

The hamlet of Claxton is recorded as Clackeston in its earliest form, confirming this to have originally been 'the *tun* or farmstead of a man called Clac'. Cloff Bridge comes from *cloh* 'a deep valley', we would normally expect to see this name as Clough. Micklemire Lane is from *mickle myrr*, 'the larger muddy place'.

Parkhursts Hospital was founded by Dormer Parkhurst, master of Greatham Hospital. Pudding Nook points to this as having 'clayey, muddy soil'. From Old English *pynd fald* comes Pinfold, a name common to many places for it speaks of where stray animals were held awaiting collection by the owner upon payment of a fine. Nobles is a field name which also mentions money, the noble being the rental payment, an amount equal to six shillings and eight pence. If this seems an odd amount, it may make more sense if given in the current decimal coinage equal to 33.3 pence or one-third of a pound. Note the name appears in the plural today, however this is more likely due to it being misinterpreted as a personal name, where the 's' is possessive, rather than a number of coins.

Great Lumley

A name of Old English derivation which is recorded as Lummalea around the middle of the eleventh century. Coming from *lumm leah* this is 'the woodland clearing by the pools'.

The Dog and Gun public house shows an image of the man and his gun dog.

Great Stainton

Listed as Staninctona in 1091, Steinintune in 1250, Staynton in 1284 and as Staynton in Strata in 1312. This last record is also seen as Little Stainton and Stainton-in-the-Street. These places, with self-explanatory additions of Great and Little, come from *straet tun* 'the farmstead on the Roman road'. The final record of Stainton-in-the-Street also features the element *straet*, showing this was named late, with the original meaning of the place name unknown.

Elstob is a minor name from *ellern stobb*, showing it was marked out by 'an elder tree stump'. Hylton is seen as Hilton in 1312, Helton in 1335, and Hylton for the first time in 1539, the name clearly referring to 'the hill by or with a farmstead'.

H

Hamsterley

Hamsterley is difficult to define for aside from the record of Hamsteleie at the end of the twelfth century, every other record is exactly as it appears today. While the suffix is clearly *leah*, the first element may be a personal name or the word *hamster*. In the first instance the name would be Frisian and give 'the woodland clearing of a man called Hamstra', while *hamster* is of German beginnings and speaks of 'the woodland clearing infested by the corn weevil'.

Redford is a common name, the early forms of Roteford and Rohford in the fourteenth century show this to be different from the majority in being 'the rotten ford', that is a soft bed making it boggy. Cranerow features a surname, this describing 'the row of houses associated with the Crane family'. Hoppyland is a difficult name, the suffix is probably *land* referring to 'cultivated land' and would follow a personal name but no examples come readily to mind from the few early forms available. Snapegate is recorded as Snaypesget in 1382, here Old English *gest* follows the dialect word *sneip* or *sneyp* (originally from Old Scandinavian *sneypa*). The first element means 'to be hard on, snub, or rebuke' and, while the suffix is uncertain, seems to be speaking of land which was notorious for repeatedly refusing to respond to any fertilisation or cultivation techniques employed by the farmers.

Mayland Lea is first seen in a document dated 1382 as Mayland, Lea being a later addition for 'pasture'. The basic name probably represents Old English *maegthe land* which, while the first element may be used as a personal name, actually describes 'the agricultural land of a woman'. The first element is also seen in other English place names, such as Mayfield in Sussex and Maghull in Lancashire, where some sources have suggested this to be *maeg*. While *maeg* or 'virgin' is well known, it is only used in literature. Furthermore as personal names in a place name are almost exclusively used posthumously, the idea of a matriarch of a society being a virgin makes no sense, hence *maegthe* is given.

Hart

The village of Hart which must share its origin with the larger town and port of Hartlepool, almost certainly transferred here and thus not really having an etymology in the normal sense. The place is claimed as the birthplace of Robert the Bruce, King of Scotland.

Locally we find Dalton Piercy, a name describing 'the valley settlement' and one associated with William de Percy by the thirteenth century. Bogle Beck earns its name from the dialect term *bogle* meaning 'a phantom, hobgoblin'. Nelson began life as 'the farmstead of a man called Niel'. Pudding Street tells us it was muddy. Nesbit comes from *nesu bite* 'the nose-shaped bit of land'. What was 'the outlying farmstead of the Bulner family', associated with Ralph de Bulmer by 1312, appears on today's maps as Thorpe Bulmer. Shortcake Hill has been suggested as referring to shape, however it is more likely to refer to the crumbly texture of the soil. While Thorston could be either 'the farmstead of a man called Thori or Thuri'. The personal name featured in Morleston is probably of Old French derivation, this being 'the farmstead of a man called Morel'.

Hartlepool

A place listed as Herterpol around the end of the twelfth century, this name comes from Old English *heorot eg pol* and describes 'the pool or bay near the peninsula frequented by stags'.

Street names begin with Alma Street, named after the battle which was fought on 20th September, 1854, considered the first battle of the Crimean War it is often seen as a pub name. Baltic Street was named for it being associated with those principally trading with the Baltic provinces of Estonia, Livonia and Courland. Cleveland Street commemorates the Marquis of Cleveland, who served as mayor in 1831. Another mayor was William Vollum, a ship owner who served five terms as mayor and who had Vollum Road named after him. Prissick Street was named in 1841 to mark the work of philanthropists Henry and Elizabeth Prissick. Stripe is an old street name, taken from a field name referring to a long, narrow strip of land.

The hamlet of Brierton appeared as Brerton and Brereton in the fourteenth century, while quite recently the name is recorded with the alternative spelling of Brearton. The name is of Old English origin in *braer tun* 'the farmstead of the brambles'. Stranton comes from Old English *strand tun* 'the place on the shore'. Throston has been recorded since the fourteenth century, although 'the farmstead of a man called Thor' will have existed since shortly after Roman times. Trickley is derived from a little used English word *trickle* describing 'the farm of sheep dung'.

On the coast is the name of Eden, seen as Geodene, Iodene and Yoden in early records, this speaks of 'the valley of Goda'. From *naes byht* comes the name of Nesbitt and describes 'the bend at the headland'. Found as Ovetun in the twelfth century, the name of Owton is from a very common personal name and refers to 'the farmstead of a man called Offa or Ofa'.

Seaton Carew is found as Ceattun, Setona, Seton Carrowe, and Seton Karrow between the thirteenth and fourteenth centuries. Here Old English *sae tun* refers to 'the farmstead on the sea'. The addition is seen in the name of former tenant Petrus Carou, whose surname appears to be a nickname rather than a place name as there are no references as 'de Carou'. Another Old English name is Throston, a name speaking of 'the *tun* or farmstead of a man called Thurstan'. Inscar Point is a coastal feature derived from Middle English *sker* 'a rocky cliff', while further along the coast we find Maiden Bower, traditionally where a young woman was flung to her death on the rocks by her violent lover. Possibly describing 'the farmstead of a man called Neale', Nelson is first seen as Nelestune in 1196.

Harton

Listed as Heortedun in 1104, this is from Old English *heort dun* or 'the hill frequented by harts or male deer'.

Little doubting the minor name of Caudwell is from Old English for 'the cold spring or stream'.

Haswell

Recorded as Hessewella in 1131, the name is derived from Old English *haesel wella* and referring to 'the stream where hazels grow'.

Locals enjoy a drink at the Pemberton Arms, a reminder of the Pespool Estate being purchased in 1808 by John Pemberton, a barrister from York. Later the Hawthorn Estate was added to their possessions by Richard Pemberton, and his descendants further increased the size of their holdings. The Oddfellows Arms is a common name, a reference to the Independent Order of Oddfellows, a benevolent society seen in the United Kingdom since the nineteenth century.

Haughton-le-Skerne

Found around 1050 as Halhtun, the basic name comes from Old English *halh tun* and describes 'the farmstead by the nook of land'. The addition comes from the River Skerne, itself from Old Scandinavian *ski* and refers to 'the clear stream'. Skeringham must also come from the river and thus describe 'the homestead of the people of the Skerne'.

Minor names here include Coatham Mundeville, where the place 'at the cottages' was associated with the family named Mundeville or Amundaville by 1197. Greystones speaks for itself, however the term is only ever used for a marker stone. Humbleton comes from *hamol dun*, the place at 'the crooked hill'. Morton is from *mor tun* 'the farmstead near a swamp'.

Hall Garth is from *halh garth*, understood as 'a modern house'. Whiley Hill is from *wilig* meaning 'the willows', telling us it was grown here as thatching material. Whessoe is probably a nickname with *hoh* 'the spur of land of one known as Hvassi'. What began as 'the farmstead on the moor' was seen as Morton Palms after Bryan Palmes came here around 1569. From Old English *set berg* meaning 'the flat-topped hill' comes the name of Sadberge. Carcut Beck is a corruption of *cald cot* 'the cold or exposed cottages'. Street House stands on a Roman road, always referred to as a *straet* by the Saxons. With Sadberge found as Satberga in 1150, the first element here is uncertain but possible describes 'the hill with a seat-shaped top'.

The Highland Laddie may have been chosen for some association with Scotland, yet the origin is undoubtedly the song played and sung in many a home wherever a piano was once present. When the Marquis of Huntly departed with his regiment in 1799, Mrs Grant of Laggan wrote the song which began "Oh where, tell me where, is your Highland Laddie gone?"

Hawthorn

Listed as Hagethorn in 1155, here we find Old English *hagu thorn* speaking of the '(place at) the hawthorns'.

The Stapylton Arms is named after the family who were powerful landowners in the northeast of the country and can trace their line back to before the Norman Conquest.

Headlam

Here is an Old English place name recorded as Hedlum around 1190. Here Old English *haeth leah* tells us of the '(place at) the woodland clearing where heather grows'.

Hebburn

To the west of Jarrow is this settlement which is recorded as Heabrym, Hebine, Henerine, Hebern, Heabyrine, Hebbarne, and Hebbarne between the twelfth and the fifteenth century. This is from Old English *heah byrgen* and describes 'the high burial place or tumulus'.

Heighington

The earliest record of this name is as Heghyngtona in 1183, coming from Old English *ing tun* preceded by a Saxon personal name and telling us of 'the farmstead associated with a man called Heca'.

The hamlet of Coatsay is recorded as Cotes super more in the thirteenth century. This shows the modern form to be a shortened version and this name means 'the cottages on the moor'. Brakkes is from Middle English *brak*, describing the 'break or uncultivated ground between worked plots'.

The Cumby Arms remembers Captain William Pryce Cumby, who served in the French Revolutionary and Napoleonic Wars, commanding the *Bellerophon* at the Battle of Trafalgar (the ship known as the 'Billy Ruffian' by her crew). The Dog Inn features an image of a dog, a breed not readily identifiable, seated in a rocking chair with pipe in one hand and foaming tankard in the other. The sign suggests the name is a welcome, yet it may have begun as something else.

Heighington was where one of the world's first steam locomotive was constructed which used coupling rods not chains and gears in its 0-4-0 wheel arrangement, hence the naming of the Locomotion Number One public house.

Hesleden

Listed around 1050 as Heseldene, this name comes from Old English *haesel denu* and telling us of 'the valley where hazels grow'. To the east are two places which share this name, High Hesleden shows this was at greater height and Monk Hesleden was associated with the monastery.

Hardwick is a common place name, derived from *heorde wic* this is 'the specialised farm of the herdsman'. Most often *wic* is used to describe a dairy farm, and here 'herdsman' would seem to suggest cattle. However the original meaning of herdsman was one who tended the flock – if this sounds odd remember we still speak of 'herding a flock of sheep', hence 'herd' is used as a verb not a noun. Hulam comes from Old English and refers to this place as being found 'at the hollows'.

The Pemberton Arms is the local pub named after the family who were major landowners in the area until the late nineteenth century.

Hett

From around 1168 comes the record of Het, a name from Old Scandinavian *hetti* or Old English *haett*, both of which describing 'the hat shaped hill'.

Hetton-le-Hole

Seen as Eppedon in 1197, with later listings of Heppedun, Heppale, Hepton, Hettone, and Hepton in Valle, all showing the original suffix was *dun* and not *tun* as the modern form suggests. Hetton-le-Hill shares this origin. Here is 'the hill of a man called Aepe', with the later addition of 'the valley'.

The local hamlet of Eppleton is another of Old English origin, listed as Appledon in 1197 and as Applynden in 1345 and referring to 'the valley where apple trees grow' from *aeppel denu*. Thickley comes from a Saxon personal name and Old English *leah* and tells of 'the woodland clearing of a man called Ticca'.

High Force

Often heard said to be the highest waterfall in England, this drop of 71 feet is over 500 feet less than Cautley Spout in Cumbria. However the sheer volume of water makes it a splendid sight, as the whole of the River Tees drops over a cliff edge created by the igneous rock forming Whin Sill, better seen as the outcrop forming much of Hadrian's Wall, falling to the eroded Carboniferous limestone at its base. As a name the High Force, there is a Low Force further downstream, is simply a dialect term influenced by the Old Scandinavian for 'waterfall'.

Holwick
Recorded as Holewyk in 1235, this name probably comes from Old English *hol wic* and describes 'the specialised farm in a hollow'.

Horden
An Old English place name from *horu denu* and meaning 'the dirty or muddy valley', the record of Horedene in 1170 is the earliest known form.

The local is the Bell, a reminder that the pub and the church were once much closer than today.

Houghton le Side
The earliest surviving record dates from around 1200 as Hoctona. The basic name comes from Old English *hoh tun* and refers to 'the farmstead near a hill spur', with the later addition from *side* 'the hill slope'.

Houghton-le-Spring
Early records just cover the basic name, seen as Hocton, Hoghton and Houghton. This comes from Old English *hoh tun* 'the farmstead by the hill spur', with the addition a reference to the Le Spring family, recorded here in the fourteenth century. Street names begin with Abbott Street, a name associated with the holy well. Anderson Square honours John Anderson, member of the District Board of Health.

Newbottle is a local name recorded as Newbotill and derived from Old English *niwe botl* and describing 'the new building'. Penshaw is a name coming from Celtic *pen* Old English *jaeg* giving 'the enclosure by the hill'. Warden Law first appears as Wardelau in the twelfth century, which is from *weard hlaw* 'the watch hill'. Shiney Row is an unusual name, for while it is certainly Old English *scien raew* 'the beautiful row', it does not follow the usual rules of toponymy. Normally we would expect the suffix to refer to a row of cottages or houses and yet there is no physical or written evidence to support this. Similarly the first element would normally be derogatory, places nearly always named by the neighbours and such humour is common in place names. Should we accept the first element really does mean 'beautiful', then it seems likely the 'row' here refers to trees or perhaps a line of marker stones.

Biddick is found as Bidich in 1190, as Bedyk in 1268, Bidykwaterville in 1339, Bedyk in 1382, Bedyk Ulkilli in 1382, and Bedic in 1442. The basic name puts together a Saxon personal name and Old English *dic* telling of 'the ditch of a man called Beda'. Note the additions in the forms of 1339, referring to its position in a meander of the River Wear, and in 1382, noting ownership by someone called Ulkell or Ulfketill. Moor House and Moorsley speak of 'the house on the moor' and 'the mound on the moor' respectively.

Penshaw is a combination of a Celtic and Old English name referring to the 'enclosure by the hill'. Eppleton is seen as Aepplingdene in 1180, Heppligdene in 1195, and Epplingden in 1311. The first record would seem to suggest *aeppel* 'apple trees' and yet to find this followed by *inga dene* is unheard of, indeed *inga* is aways connected to a personal name. As nobody would be called 'apple tree', not even as a nickname, we must assume this is a mistake and this is a diminutive of Eppa or Aeppa. Hence we define Eppleton as 'the valley of the family or followers of a man called Eppa or Aeppa'. Note also the later suggestion of *tun* is entirely down to local pronunciation, there is no doubt the original suffix was *dene*.

Herrington is of Old English derivation, the Saxon personal name followed by *tun* and speaking of 'the farmstead of a man called Hering'. Offerton has been recorded since the eleventh century, seen as Uffertun, Ufferton, and even Houghton Oufferion in 1627. This is from *tun* and a Saxon personal name describing 'the farmstead of a man called Ultfara'. Warden Law appears as Wardona in 1104, this being 'the watch-out hill'. Wooden speaks of 'the hill of the wolf', although whether this refers to the animal or is used as a personal name is unclear.

Howden-le-Wear

Howden is recorded as Holden at the end of the thirteenth century, showing the modern form to be a corruption and probably influenced by the Howden in nearby Yorkshire. Here the name comes from Old English for 'the hollow or deep valley'. The addition distinguishes it from its namesake and refers to the river meaning 'the winding stream', with the French

Fingerpost at Howden-le-Wear

definite article remaining despite the loss of the preposition.

At a junction to the northwest is Fir Tree, which has no recognisable early forms. This is unlikely to be anything to do with fir trees, this probably represents Old English *fergen* referring to 'the wooded hill' with the later mistaken addition of 'tree'.

More trees are seen in the name of the local pub, the Plantation Inn. The Australian is clearly pointing to an inn-keeper or owner who came from that southern hemisphere continent.

Hunderthwaite

Domesday lists this name as Hundredestoit in 1086. Here Old Scandinavian *thveit* is preceded by a personal name and describes 'the clearing of a man called Hunrothr'.

Hunstanworth

Found in a document dated 1183 as Hunstanwortha, here a Saxon personal name is followed by Old English *worth* and telling us it was 'the enclosure of a man called Hunstan'.

Nuckton is a minor place name found as Knokeden in the twelfth century. Clearly the ending has changed from *dun* to *tun*, the first element is probably *cnoc* and thus 'the farmstead at the hill'. Allenshiel describes 'the huts of the summer pastures associated with a man called Aleyn', and for once we can identify the man for this is a Middle English name and one which refers to Alan the Marshall, the former landowner.

Hunwick

The middle eleventh century listing as Hunewic shows this comes from a Saxon personal name and Old English *wic* and referring to 'Huna's specialised farm'.

Helmington Row sign

Helmington is a local name describing 'the farmstead of the family or followers of a man called Helm', where the Saxon personal name is followed by Old English *inga tun*.

Pubs here include the Joiners Arms, the coat of arms granted in 1571 and probably signifying a landlord or owner who was associated with this trade. The Quarry Burn is a description of, not the pub itself, but of the destination of Quarry Burn Lane which gave it a name.

Hurworth

The record from 1158 gives this name as Hurdewurda, this is from Old English *hurth worth* and describes 'the enclosure made with hurdles'. Nearby Hurworth-on-Tees adds the name of the river, itself discussed under its own entry.

Street names include Cree Beck, *crew* being a dialect term meaning 'a pen, fold, sty, or enclosure'. High Rawcliffe takes a place name describing 'the high red cliff'. Newbus is a corruption of *niwe hus* 'the newer house'.

Neasham is a minor name coming from *nesu ham*, describing 'the nose-shaped homestead'. Greenland Plantation is a remoteness name, a term describing the farthest flung corner of the parish. Such names are often datable, for they would have been in the news around this time, perhaps newly discovered or settled. Another common theme is, rather surprisingly, the derogatory name, which also goes to show how places were named by the neighbours, to those who lived at Hunger Hill it would have been simply 'home'. The Otter and Fish is named for the mammal and its prey found in the nearby river.

The Emerson Arms takes the name of the family who lived at Chatwin House and were related to the American philosopher Ralph Waldo Emerson. Another famous Emerson was born to the village schoolmaster and his wife in 1701. Named William he was known locally for being a most eccentric, uncouth and surly individual. However his genius in mathematics spread far and wide, indeed he was invited to become a member of the Royal Society. He declined the honour, none too politely, objecting to having to pay his own membership fees.

At Dalton-on-Tees, the 'farmstead in the dale', we find the Chequers Inn. This is one of the oldest pub names in England and can be traced back to the Roman occupation when it was used to indicate a board game similar to draughts being played within. Later it came to represent a moneyer, the term is still seen today for the British government department responsible for the country's finances is still known as the Exchequer.

Hutton Henry

Hutton is a common place name, found in the middle of the eleventh century as Hoton and derived from Old English *hoh tun* which speaks of this as 'the farmstead on or near the hill spur'. The addition, to distinguish it from the following entry, is seen from the fourteenth century when the manor was held by Henry de Essh.

Catlaw Hall is from *hlaw* with a personal name and describes 'the mound of a man called Catta'.

Hutton Magna

Another name from Old English *hoh tun*, again describing this as 'the farmstead on or near a hill spur'. Here the addition is from Latin *magna* meaning simply 'great'. The name is recorded as Magna Hoton in 1157.

I

Ingleton

A name recorded as Ingeltun around 1050, which comes from a Scandinavian personal name and Old English *tun*. This name speaks of 'the farmstead of a man called Ingeld'.

Iveston

With the earliest known record dating from 1183 as Ivestan, this name is certainly derived from a Saxon personal name and Old English *stan* which is seen as 'the boundary stone of a man called Ufa'.

Stockerley, recorded as Stokerley in 1382, features the suffix *leah* to speak of 'the woodland clearing associated with a stocker'. Whilst today we see this as a family name and not a trade, originally this was the description of a man who grubbed up *stoccs* or tree stumps. This is not to say the *leah* was one of tree stumps, a *leah* is a natural clearing, one where underbrush covers an area where trees will not root. Here the *leah* was where the man lived.

J

Jarrow

Records of this name include Girvum, Girwa, Girvi, Gerwin, Jarue, Gyrwe, Gryuum, Jarwe, and finally Jarrow. There is an earlier record from 757 as Donaemuthe and, as Jarrow stands at the 'mouth of the River Don', may well have referred to this town. However the modern name is best seen in the record of Gryuum from around 730, a name which probably speaks of the '(place of) the fenland people'. This would reflect the tribal name Gyrwe, itself derived from Old English *gyr* 'mud, marsh'.

Locally we find Hedworth, a name from Old English *haeth worth* to refer to 'the enclosure on the heathland'. Monkton still largely speaks for itself, a former monastical property this is 'the farmstead of the monks'. Felling, often referred to locally as The Felling, is indeed where trees were felled for agricultural use. Follingsby's early forms show this is not manorial but likely from *full litha by* 'the farmstead fully provided with troops'. Brockley Whins describes 'the clearing where furze grows frequented by badgers'. Harton is recorded since the twelfth century, this certainly represents Old English *heort tun* which would give 'the farmstead frequented by harts', however it should also be noted Heorta is a personal name derived from the same word.

Headworth has two possible meanings and it is difficult to decide between *haeth worth* 'the enclosure of the heath' and 'the enclosure of a man called Haethe'. Wardley links Old English *leah* with a Saxon personal name and speaks of 'the woodland clearing of a man called Wearda'. Westoe began life with the suffix *stow*, this speaking of 'the place of a man called Wifa'.

K

Kelloe
Found as Kelfau around 1170, this is derived from Old English *celf hlaw* which describes 'the hill where calves graze'.

Kibblesworth
Records of this name include Kibbeswurhya, Hyblesworth, Kibleswig and Kebillsworth. Here a Saxon personal name and Old English *worth* speak of 'the enclosure of a man called Cybba'.

Killerby
Here is a name from Old Scandinavian *by* preceded by a personal name. This tells us the place was described as 'the farmstead of a man called Ketilfrothr', the place being recorded as Culuerdebi in 1091.

Kimblesworth
Listed as Kymliswrth during the thirteenth century, this name features a Saxon personal name and Old English *worth* which speaks of 'the enclosure of a man called Synehelm'.

Locally we find Findon Hill, three syllables where *fin dun hyll* refers to 'the heap hill, hill', the 'heap' most likely a barrow but possibly a cairn.

Fingerpost at Killerby

Knitsley

Records of this name include Knyhtheley in 1303, showing this comes from Old English *cniht leah* and describes 'the woodland clearing of the retainers'.

Ivesley appears as Ivesleyburdon in 1382 and refers to 'the woodland clearing of a man called Ifa or Ivo'.

L

Lanchester

A record of Langecestr from 1196 probably points to this place name coming from Old English *lang ceaster* and meaning 'the long Roman stronghold', appearing so because it is aligned along the road providing access.

Minor names include Collierly, which has no connection with coal mining but is a corruption of 'the *leah* or woodland clearing or a man called Ceol or Ceolwulf'. Hamsteels is Old English *ham steall* 'the homestead with shelters or stalls for cattle'. Kyo is a much shortened 'hill or mound of a man called Cy or Cyne', the suffix being Old English *hlaw*. Westoe is found as Wywestoue in 1195 and as Wiuestoue in 1335, these are from Old English *hoh* and a Saxon personal name and speak of 'the spur of land of a man called Wife or Wifa'.

Ousterley was originally Oustrefeld, the first element probably describing a plant which grows on buildings, literally referred to as a 'house tree'. Hence originally called 'the open land by the house overgrown by plants' is now 'the woodland clearing with the house overgrown by plants'. Flass comes from French *flasse* and describes the 'pools or marshy places' of the low-lying ground alongside the Deerness Brook. Billingside is the 'place alongside that associated with the family or followers of a man called Billa'. Ornsby Hill is recorded as Ormysby in 1408, here the Old Scandinavian *by* follows a personal name to tell of 'the farmstead of a man called Ormr'.

Wheatley holds no surprises when coming from *hwaet leah* or 'the woodland clearing where wheat is grown'. Pokerly is a combination of Old Scandinavian *pokker* and Old English *leah* and telling of 'the woodland clearing of the hobgoblin or demon'. Pontop is recorded as Pontehope in 1240, this coming from Old English for 'the small valley by the stream called Pont', the river name itself named after the bridge which once crossed it. Satley is a difficult name, the suffix is certainly *leah* with the possible first element *saata*, although woodland clearing with haystacks' would be a fairly unique definition.

Smallhope Burn speaks for itself as 'the stream in the small valley'. Stobbilee is found as Stubbiley in 1292, a name from Old English *stubb leah* which is among the author's favourites for it instantly describes a picture of the area in Saxon times as 'the woodland clearing marked by tree stumps'. Migley is either from Old English *micga leia* or the first element is the dialect word *migg*, both speaking of 'the manure field'. This would be one well-manured or where such was held awaiting spreading as fertiliser. Bushblades is an unusual name derived from *birse blade* and understood as 'the place where bristly blades grow'. Colepike Hall has two elements, *col pightle* describing 'a corner of land where open-cast coal mining took place'. Similarly Collierley speaks of 'the clearing where coal was mined'. Esp Green tells us it was marked by 'the aspen trees'.

From *faugh side* comes Fawside or 'the variegated land on the side of the hill', a reference to colour. Greencroft tells us this small holding was left to lie fallow, being 'green' and not turned by the plough. Healeyfield describes 'the *feld* or open land of a man called Aleyn'. Hurbuck is seen as Hurthebuck in 1303 and Hurtebuckside in 1312, this is almost certainly Old Scandinavian *hurtharbak* describing 'the back of the door' or 'the space behind it'. While this is undoubtedly used in a topographical sense, just as it does in three similar examples in Iceland and several others in Scandinavia, what it describes is uncertain. Warland can only be 'the cultivated land of a man called Wra', this being a pet form of a number of Saxon personal names. Wooley is recorded as Ulflawe in 1260, this being 'the woodland clearing frequented by wolves'.

The Roman fort here was known as *Longovicium*. It began as an interim stage on Dere Street, the main road linking *Eboracum*, modern York, with Hadrian's Wall. As a place name *Longovicium* connects the British or Celtic *longo* meaning 'ship' with Latin *vicium* 'a settlement'. This is a tantalising hint as to the origin of the forces first stationed here for clearly there is no ocean here for a ship. While a slab found here shows this was built by the Legio XX Victrix (the Twentieth legion), this does not mean they were the first stationed here. Probably a reference to a legion who were stationed here as either had been or still were a part of the *Classis Britannica*, the 'British fleet'. This being stationed here as part of the defence of this part of the Roman Empire. Alternatively, this may show the legion encountered troubles *en route* to these shores.

Langley Park

A fairly common name and one which is always from Old English *lang leah* 'the long woodland clearing'. Clearly whether such is long or broad depends upon the position of the viewer and, as it is clear that most places are named by the neighbours and not by the residents, it gives some clue as to the location of those who named it. This name is recorded as Langeleye in 1232.

Charlaw Moor features the suffix *hlaw* and speaks of 'the hill of a man called Ceorra'.

Langton

Found as Langadun around the middle of the eleventh century, this name comes from Old English *lang dun* and describes the nearby 'long hill'.

Lartington

Documented as Lyrtingtun in the middle of the eleventh century, this name comes from a Saxon personal name and Old English *ing tun*. Together they inform us it was known as 'the farmstead associated with a man called Lyrti'.

Leadgate

A name derived from Old English *hlid geat*, this refers to this place having 'a swing gate'. While a hinged gate today is the norm, originally a *geat* was simply a gap providing access to whatever lay beyond. The earliest known record of this name is as Lidgate in 1590.

Pubs here include the Golden Lion, an heraldic image which here would refer to the influential and powerful Percy family, Dukes of Northumberland.

Leamside

Records of this name begin with Le Leme in 1365 and Le Lemside in 1380. These records clearly show the Old French definite article, which can thus be discounted. The basic name comes from a Celtic river name referring to either the '(place) alongside the elm river' or '(place) alongside the marshy river'.

Minor names include Cocken, a difficult name for any suffix is long lost and what survives seems to be a personal name with the Middle English -en telling us it was the place 'of a man named Coc or Cocca'.

Long Newton

The only early record dates from 1260 and appears as Lang Newton. Here are three Old English words *lang niwe tun* referring to 'the long new farmstead'. Note this is referred to as 'long', this is of course relative, for if approached from another side it would appear 'wide'. Hence we know the early settlement was aligned along the main road running through here and doubtless named by another settlement found further along that route.

Londonderry Cottage was named after the powerful local landowners, the Marquesses of Londonderry. Similarly the Vane Arms Hotel, which was named after the family surname and not the title.

Low Dinsdale

Recorded as Ditneshall around 1185, this name tells us it was 'the nook of land belonging to Deighton (in Yorkshire)'. Here the name of the neighbouring manor precedes Old English *halh*.

Local names include High Stodhoe and Low Stodhoe, sharing an origin of *stod hoh* and describing 'the heel of land with a stud farm'; Bath Villa tells us there were cold baths available, probably open to the public; and the Fighting Cocks, while a common enough pub name is also associated with one Elisha Cocks in 1840 and thus is more likely inspired by the person rather than the barbaric sport.

Lowes, Forest of

A name virtually unchanged since the early fourteenth century. It comes from the plural of *luh* meaning 'loughs' or 'lakes'.

Ludworth

This name comes from a Saxon personal name and Old English *worth* and speaks of this as 'the enclosure of a man called Luda'.

Lynesack

Listed as Linesak, Lynsak and Lynesak in the fourteenth century, this name comes from one of two sources. Either this is Old English *hlind ac* or the '(place of) the limes and oak trees' or the '(place of) the oak tree of a man called Lin or Lind', if the first element is a Saxon personal name.

Minor names include Copley, comprised of a Saxon personal name and Old English *leah* and speaking of 'the woodland clearing of a man called Coppa'. Pooltree is self-explanatory as 'the tree by the pool'. Wham comes from Old Scandinavian *hvammr*, a very specific description of 'a valley with high lands all about but with a single opening to one side'.

M

Medomsley

Here is a name found as Medomesley, Medomsle, and Madmesl in the twelfth and thirteenth centuries. This comes from a Saxon personal name and Old English *leah* and tells us this was once 'the woodland clearing of a man called Maethelhelm'.

Locally we find the name of Bradley, a hamlet with a very common name meaning 'the broad woodland clearing' from Old English *brad leah*.

As the second most common pub name in the land, the Royal Oak is clearly of some importance. Indeed the name harks back to the seventeenth century when King Charles II, fleeing from defeat at the Battle of Worcester, hid in an oak tree at Shifnal in Shropshire with his aide Colonel Carless. To mark the event, at the Restoration of the Monarchy, the king's birthday of the 29th May was named Royal Oak Day and over five hundred pubs commemorate this famous event. From the same era comes the Hat and Feather, featuring the distinct headwear of Prince Rupert and the so-called Cavaliers. The Miners Arms is clearly a reminder of this dangerous occupation which was so important to the area.

Merrington

This name is found as Maerintun around 1085 and as Kyrke Merington in 1331. The basic name comes from a Saxon personal name with Old English *ing tun* and meaning 'the farmstead associated with a man called Maera'. The addition is Old Scandinavian *kirkja* or 'church'.

Listed as Maerintun in 1125, the same year we find Meringtonas, and Merringtun around the end of that century. Here a Saxon personal name precedes Old English *inga tun* to tell of 'the farmstead of the family or followers of a man called Maera'.

Minor names include Ferryhill, from Old English *firgen hyll* or 'the hill covered with ferns'. Hett seems to be a dialect form of 'hat' and a description of some local feature said to resemble some form of headgear. Middlestone is found as Malderstayn in 1366, the suffix *stayn* follows an uncertain first element having two equally possible Old Scandinavian origins. This may represent *malar stayn* 'the place of pebble-sized stones', although this is also *meldr* (also seen in the dialect term *melder*) referring to 'ground corn' and understood to mean 'grinding stones'.

Mickleton

Domesday records this name as Micleton in 1086, the name coming from Old English *micel tun* and referring to 'the large farmstead'.

At the beginning of the seventeenth century a pub name appeared which was to gain instant popularity, unsurprising considering the Rose and Crown showed allegiance to both the monarchy and the nation.

Middleton in Teesdale

A very common basic place name, hence the addition. Found as Middeltun around 1164, this comes from Old English *middel tun* and refers to 'the middle farmstead', ie between two others. Here the addition refers to a river name meaning 'the surging one' with Old Scandinavian *dalr* and informing us it is to be found 'in the valley of the River Tees'.

Locally we find Vallence Loo, the name transferred here via a former landholder Hamund de Valines. First seen as Valineslieu, this adds the Old French *lieu* 'place' to the surname. Named after the family who owned much of the land south of the Tees, the Strathmore Arms stands on a cul-de-sac leading to their former manor house. The Foresters is named to show a meeting place of the Ancient Order of Foresters, a friendly society with 'lodges' on both sides of the Atlantic.

Middleton St George

Recorded as Middlinton in 1238, this second example of 'the middle farmstead' is again from Old English *middel tun*. The addition here refers to the dedication of the church.

Locally we find Dinsdale, listed as Ditleshal in 1197, Dyteneshall in 1291, and as Dytensale in 1314, this would probably be 'the nook of land of a man called Dene' where the Saxon personal name is followed by Old English *halh*. Similarly Killinghall describes 'the nook of a land of a man called Cylla', although it probably came here via a family who are recorded here from 1416 until 1587, then returning in 1762, the Killinghall Arms supports this.

Over Middleton tells us it was 'the higher middle farmstead', not exactly in the middle of anything and hence borrowed its name from the parish. Poutney's Bridge is a rarity, it takes its name from Old French *pont* and describes 'the bridge (over the Tees)'. From *heorot burna* or 'the western spring frequented by harts or stags' comes the name of West Hartburn. Almora Hall has a name brought here from the district of Almora in India, hence a remnant of the days of the British Empire in India.

Fighting Cocks is a pub name, most likely influenced by the Cocks family, nineteenth century landholders. The Old Farmhouse Inn takes its name from the old building which stood here.

The pub sign, and thus its name, began as an ale stake to which a sheaf of barley was tied to show the nearby house offered refreshment to thirsty travellers. This ale stake would most often be the trunk of a tree with the lower, if not all, branches removed. It was inevitable the tree would suggest itself as a pub name, hence the origin of the Oak Tree Inn.

Mordon

Listed as Mordun around the middle of the eleventh century, there can be no doubt this name comes from Old English *mor dun* and refers to 'the hill in marshland'. Clearly *mor* would seem to be the modern 'moor', and indeed the word could be used for either, yet the topography would point to 'marsh'.

The hamlet of Elstob is found as Ellestop in the fourteenth century and as Elstobe in the sixteenth century, this name comes from Old English and describes 'the post of a man called Elle'.

Morley

A name documented exactly as the modern form as early as 1312. As stated in the previous entry Old English *mor* was, rather confusingly, used to mean both 'moor' and 'marsh'. Here the element is coupled with *leah* and thus describes 'the woodland clearing in the moor or marsh'.

Muggleswick

Recorded as Muclingewic around 1170, this features a Saxon personal name and Old English *ing wic* and tells us this was 'the specialised farmstead associated with a man called Mucel'.

Locally Steward Shiel is a reminder this was the residence of the steward of the Bishop of Durham, a shieling being 'pasture grazed in the summer only'. Similarly Carp Shield is 'the summer pastures of a man called Garpr'. Combfield House was built at 'the open land by a ridge'.

Early records of Eddys Bridge are not reliable enough for us to tell whether this is 'the bridge of a man called Aeddi or Edi' or if this refers to a woman, Edith. Hisehope is a name from Old Scandinavian *hestr* and Old English *hop* 'the small valley where horses are raised'. However we cannot entirely rule out the first element used as a personal name.

Hunterley Hill is recorded as Hunterlaw in 1342, showing the modern 'Hill' is superfluous as the original suffix was not *leah* but *hlaw* and thus the name already describes 'the hill of the hunter'.

Murton

Found as Mortun in a document dated 1155, this comes from Old English *mor tun* and tells us it was 'the farmstead in moorland or marshy ground'.

The Colliery Inn is a reminder of the importance of mining to the area, while the Village Inn is a self-explanatory and somewhat unimaginative name.

N

Neasham

A name recorded as Nesham in 1158, this name refers to the 'homestead by the projecting piece of land' and it does stand in a bend of the river. This comes from Old English *neosu ham*.

Here Sockburn is recorded from the late eighth century as Soccabyrig, Socceburg, and Sochasburg. This would point to Old English *socu burh* and refer to 'the borough of the judiciary'. Not until 1291 do we see the suffix as *burna* with Sokeburn. While there is undoubtedly a small stream or spring here, the change seems to have been a written error which stuck.

Newbiggin

The earliest surviving record of this name dates from 1316, a document where the name appears as Neubigging. This comes from Old English *niwe* and Middle English *bigging* and describes 'the new building'.

Newton

While it seems to come from Old English *niwe tun* and meaning 'the new or newer farmstead', this is not what it seems but is indeed a 'new town' which takes the name of nearby Aycliffe which is discussed under its own entry and refers to 'the oak trees at the woodland clearing'.

Newton Aycliffe

Unlike other Newtons this is indeed 'the new town', a name added to that of the local Aycliffe, itself from *ac leah* 'the woodland clearing among or by the oak trees'. Most Newtons are derived from *niwe tun*, literally 'the new farmstead' but which should be seen as 'newer farmstead' as the names have existed for several centuries and can hardly be considered 'new'.

Street names of the town do not exist, for there are none, simply roads, closes, avenues, walks, etc. Religious references are found in bishops and saints associated with Durham Cathedral. These include Biscop Crescent, St Aidan's Walk and Van Mildert Road. Aldwyn Walk took the name of Aldwyn, eleventh century Archdeacon of Durham. Alington Road was named to honour the early twentieth century Dean of Durham, Headmaster of Eton College and author of detective stories Cyril Alington. A more famous author and Christian is remembered by Bede Crescent, this the man styled 'the Father of English History', the Venerable Bede.

Educational and agricultural reforms were as much the reasons for the naming of Barrington Road, although Barrington Shute was also Bishop of Durham from 1791 to 1826. Anthony Bek was another Bishop of Durham also remembered for other exploits, such as leading the English against the Scots and as King of the Isle of Man, and still seen here in Bek Walk. Butler Road remembers the author of *Fifteen Sermons* and *Analogy of Religion*, acclaimed philosophical works by former Bishop of Durham, Joseph Butler.

Bury Road remembers another former Bishop of Durham, Richard de Bury whose work *Phiobiblon* published in the fourteenth century would have been on the shelves of the first lending library in the country, established by Bishop Bury in 1340. Crewe Road remembers Nathanial, Lord Crewe, whose term as Bishop of Durham lasted for 47 years seeing service under five monarchs: Charles II, James II, William III and Mary II, Anne, and George I. Creighton Road recalls Mandell Creighton, educated at Durham Grammar School and was appointed Bishop of Peterborough in 1891 and on to London in 1896, meantime he penned one of the most important historical works of its kind in *Creighton's History of the Papacy*.

Carileph Close recalls William de St Carileph, eleventh century priest who began the building of the present Durham Cathedral in 1093. Edward Chandler wrote *Defiance of Christianity* on his way up to becoming Bishop of Durham and is remembered by Chandler Close. Cosin Close remember John Cosin, seventeenth century Bishop of Durham who spent much of his own money on charities and Durham Cathedral. St Cuthbert's Way recalls the man who is patron saint of County Durham and former Bishop of Lindisfarne.

Dykes Walk takes the name of John Dykes, vicar of St Oswald's church and composer of several hymns, including *Nearer My God to Thee*. Founder of a monastery at Ebchester and daughter of

Aethelfrith, King of Northumbria, St Ebba is remembered by Ebba Close. Faber Close recalls Frederick William Faber, who penned many hymns during his time in, first, the Church of England and, later, the Roman Catholic faith. St Godric, seen in the name of St Godrics Road, recalls the hermit who apparently had great power over wild animals and who died in 1170 at the reported age of 105.

Bernard Gilpin, rector of Houghton le Spring in the sixteenth century, worked tirelessly for charities and to found his local grammar school earning him the soubriquet Apostle of the North and the naming of Gilpin Road. Greenwell Road remembers William Greenwell, nineteenth century canon of Durham Cathedral who did much of the groundwork in identifying the early history of County Durham. Hatfield Road honours Thomas Hatfield, Bishop of Durham, King's Secretary and Keeper of the Privy Seal in the fourteenth century.

Heild Close is after Heild, sometimes given as Hild, a seventh century Saxon who became abbess of Hartlepool in 649 and of Whitby eight years later. Henson Grove recalls Herbert Hensley Henson, twentieth century bishop of Durham noted for his important theological works. Kellawe Place recalls Richard Kellaw, sometime given as Kelloe, the early fourteenth century bishop of Durham who enlarged Stockton Castle and tried to put an end to the border troubles with Scotland which put even greater strains on the quality of life for the already suffering locals.

Langley Road recalls Thomas de Langley, fifteenth century bishop of Durham who helped negotiate a truce with France and peace with Scotland. Joseph Lightfoot, theologian and writer of some renown in the nineteenth century, gave a name to Lightfoot Road. Markham Place recalls William Markham, eighteenth century Dean of Durham and later Archbishop of York. A similar career was seen for John Moore, Moore Lane named after the former Dean of Durham and later Archbishop of Canterbury.

Moule Close recalls Handley Can Glyn Moule, who served 20 years as Bishop of Durham at the beginning of the twentieth century. Paulinus Road takes the name of Paulinus, a man credited with bringing Christianity to County Durham in the early seventh century. Henry Philpotts was an early seventeenth century Dean of Durham and later Bishop of Exeter who gave a name to Philpotts Walk. Pudsey Walk remembers Hugh de Pudsey, an early twelfth century Bishop of Durham and Earl of Northumberland who influenced many of County Durham's buildings.

Ross Walk took the name of John Ross, former Dean of Durham and later Bishop of Exeter. Secker Place recalls Thomas Secker, rector of Houghton le Spring who went on to become Archbishop of Canterbury. Walter Skirlaw, Bishop of Durham from 1388 to 1406, gave a name to Skirlaw Road and helped re-build the first bridge across the Wear at Durham. Tunstall Road comes from Cuthbert Tunstall, Bishop of London and later Bishop of Durham. Walcher Road recalls Walcher, appointed Bishop of Durham by William the Conqueror in the late eleventh century. Flambard Walk remembers the twelfth century Bishop of Durham Ralph Flambard, who owed most of his success to William II and soon fell out of favour when Henry I acceded to the throne.

Warburton Close takes the name of William Warburton, prebendary of Durham Cathedral and later Bishop of Gloucester in the eighteenth century. Wheeler Green is named after Sir George Wheeler, rector of Houghton le Spring in the early eighteenth century. Wiseman Walk recalls Cardinal Wiseman, the first Archbishop of Westminster who was educated in Durham. Espin Walk recalls Thomas Espin, rector of Tow Law and also an accomplished musician, geologist, botanist, and renowned astronomer.

Influential families have given their name to Bellasis Close, Beveridge Way, Bowes Road, Eden Road, Shafto Way and Silkin Way. George Allan is remembered by Allan Way and has played an important role in the history of the town. Having joined his father as a solicitor, he was soon able to devote his life to research when he married a Yorkshire heiress. His research into genealogy, heraldry and natural history were well documented by his own printing press. Furthermore the museum he set up contained a number of important specimens, including those brought back by Captain James Cook on his voyages to the South Seas.

Anne Swyft Road was named after the wife of the sixteenth century Chancellor of Durham. While she is remembered as the first woman ever to walk inside the college, it was for her petitioning of James I in 1605 to erect a grammar school in Newton Aycliffe and for her legacy to the poor of the parish that she was honoured with the naming of the road. Arrowsmith Square recalls Aaron Arrowsmith, who trained as a mathematician and, cheated out of his inheritance by his stepfather, went to London and earned fame as a prolific cartographer of considerable skill. Thomas Bewick earned fame as an accomplished wood engraver before being immortalised by the name of Bewick

Crescent. Crowley Place takes the name of Sir Ambrose Crowley, whose early eighteenth century ironworks was the focus of much of his work for the community. Carr Place remembers banker Ralph Carr, who opened the provincial bank in Newcastle-upon-Tyne in 1755, the first of its kind in England.

Michael Denham gave a name to Denham Place, a collector of antiquities who is buried at Piercebridge. Dent Walk remembers Joseph Dent, founder of the publishers who produced, among other things, *Everyman's Library*. Dixon Road remembers Jeremiah Dixon, a name unlikely to ring any bells until this geographer is mentioned alongside his colleague Charles Mason, together they produced what became known as the Mason-Dixon Line across the United States of America separating the Slave States in the south from the Free States to the north.

Doxford Close recalls the man who, along with his father, established a ship-building business. Sir Theodore Doxford was also a member of the British Corporation for the Survey and Register of Shipping, a member of Sunderland Town Council and MP for Sunderland. Ewbank Close remembers J W Ewbank, member of the Royal Scottish Academy whose art career amounted to just two pictures and died in poverty. Fowler Road recalls John Fowler, remembered as the inventor of the steam plough.

Henry Greathead invented the lifeboat as we know it today, Greathead Crescent was named in his honour. Greville Way is named after the botanist Robert Kaye Greville. Hackworth Close recalls Timothy Hackworth, railway pioneer who did much to improve Stephenson's work on the Stockton and Darlington. Hartley Road was after Sir Charles Augustus Hartley, a civil engineer whose career improved the navigation and ports along the lower reaches of the River Danube.

Sir James Wycliffe gave a name to Wycliffe Close, this political historian was among the delegation to the peace conference at the end of the First World War. Henderson Road recalls Arthur Henderson, former labour leader and statesman. Humphrey Close was named after Thomas Humphrey, a famous nineteenth century maker of time pieces. Hunter Road remembers Sir George Burton Hunter, a shipbuilder who is remembered for bringing the building of the *Mauretania* to the region.

Sir Joseph William Isherwood, a ship designer who improved methods and had Isherwood Close named after him. Jackson Place recalls Ralph Ward Jackson, who is rightly recognised as the founder of West Hartlepool after selecting the area where the new railway would come and meet the new docks. Kemble Green recalls Stephen Kemble, a member of a highly successful acting family and brother of the best known of them all, Sarah Siddons.

Lamb Close recalls Robert Lamb, a Durham man who wrote *A History of Chess*. Liddell Close took the name of Henry George Liddell, Dean of Christ Church College, Oxford and publisher of a Greek to English dictionary. However it is his daughter who has the greater claim to fame, for Alice Liddell was the eponymous character in the works of Lewis Carroll. Another wordsmith, John Lingard, who wrote *History of the Anglo-Saxon Church* had Lingard Walk named after him.

Mellanby Crescent was named after John Mellanby, a man who rose to become Professor of Physiology in both London and Cambridge. Merz Road was named after the electrical engineer Charles Hersterman Merz and the man who designed the first electrical power station in Britain and also worked abroad. Mills Close recalls shipping engineer Sir William Mills, his was the first factory producing aluminium in Britain and also produced hand grenades at a factory in Birmingham during the First World War.

Morton Walk recalls playwright Thomas Morton, best known for *Speed the Plough*. The famous Palmer's Shipbuilding Yards at Jarrow were named after Sir Charles Mark Palmer, as was Palmer Road. Pease Way remembers Edward Pease, a railway promoter on the first passenger line from Stockton to Darlington. Priestman Road is after Sir John Priestman, a leader among Sunderland's shipbuilders. Raine Walk recalls James Raine, nineteenth century antiquarian and historian of Durham.

Ritson Road was named after the writer Joseph Ritson, who penned numerous works including *Observations on Johnson* and *Steven's Edition of Shakespeare*. Sanderson Close recalls Frederick William Sanderson, headmaster of Oundle School. Seton Walk recalls Ernest Thompson Seton, a Canadian naturalist and wildlife painter and author. Sharp Road recalls antiquarian and historian Sir Cuthbert Sharp who also served as Mayor of Hartlepool for many years.

Furniture designer and manufacturer Thomas Sheraton in remembered by Sheraton Road. Locally born musican and composer William Shield is remembered by Shield Walk. Smith Walk seems a little ordinary for Jane Elizabeth Smith whose talents for languages meant she was able to teach French, Italian, Latin, German, Syrian,

Arabic, Persian and Hebrew along with her native English, and also geometry, algebra and music. No surprise her hectic schedule resulted in death from exhaustion at the age of 29.

Nineteenth century artist Clarkson Stanfield is remembered by Stanfield Road. Journalist W. T. Stead, editor of the *Northern Echo*, is honoured by the naming of Stead Close. He died when his vessel, *RMS Titanic*, was hit by an iceberg. Taylor Walk remembers another editor, this time Tom Taylor was editor of *Punch*. No surprise to find Stephenson Way was named to mark the railway pioneer George Stephenson. County Durham's most celebrated historian Robert Surtees is recalled by Surtees Walk.

Sir Joseph Swan, inventor of the incandescent lamp, was born in Sunderland and had Swan Walk named after him. Chemist John Walker, who gave his name to Walker Lane, was probably honoured for inventing the one item which never works until it strikes, the match. Wallas Road remembers Graham Wallas, political psychologist and founder of the London School of Economics and Political Science.

Barrister and MP for Stockton and Thornaby Sir Bertrand Watson gave his name to Watson Road. Labour politician William Whiteley is honoured by the name of Whiteley Grove. Ellen Wilkinson, after whom Wilkinson Road was named, did much work for trade unions, women's suffrage, and represented Jarrow as MP. Wright Close remembers eighteenth century mathematician and astronomer Thomas Wright.

Auckland Place recalls George Eden, Earl of Auckland and former Governor General of India after whom the New Zealand city was named. Bradford Close took its inspiration from Brigadier General Roland Boys Bradford VC MC, who was killed at the Battle of Cambrai in November 1917 when the youngest Brigadier General in the British Army, he was just 25. Nathan Brunton gave his name to Brunton Walk, former Vice-Admiral in the Royal Navy whose first taste of life at sea was on a collier vessel operating out of the northeast.

Bertie Road honours Sir Thomas Bertie, who commanded a vessel at the Battle of Copenhagen under Nelson, later promoted to Rear-Admiral and thereafter Admiral. Christopher Walk is after William Christopher, commodore of the Hudson Bay Company and who sailed with Captain James Cook on his third and final voyage. Captain William Cumby commanded *HMS Bellerophon* at the Battle of Trafalgar and was still in command when taking the surrender of Napoleon and again when conveying the French emperor to exile on St Helena. Cumby Road was named to honour his long service.

Crawford Road is after Jack Crawford, a Royal Navy man who fought at the Battle of Camperdown. With the vessels colours shot away from the mast Admiral Duncan asked for a volunteer to nail it to the broken mast. Under fire, he was wounded in the face as a result, he single-handedly did as his commander asked and has a bronze statue erected in his honour along with the road name and a silver medal.

Gort Road remembers John Surtees Prendergast Vereker, Viscount Gort. Among his military and political honours were Governor of Malta, Governor of Gibraltar, Field Marshall, and Commander in Chief of the British Expeditionary Force, all during the Second World War. Gunn Lane remembers George Ward Gunn VC MC, a Second Lieutenant awarded his Victoria Cross posthumously for his actions at Sidi Rezegh in November 1941. Hardinge Road remembers Lord Harding, a man who fought with Wellington in the Spanish Peninsula Wars, losing his hand in the process, before a career change saw him as Governor General of India and Secretary of State for War.

Havelock Close reminds us of Major General Sir Henry Havelock, who led British forces in India, including the sieges of Cawnpere and Lucknow. Kenny Walk recalls another Victoria Cross recipient, Thomas Kenny bravely tried, albeit unsuccessfully, to rescue an officer under heavy fire in France in November 1915. Marshall Road remembers Sir William Raine Marshall, First World War Lieutenant General who commanded the forces against the Turkish Army, forcing their eventual surrender in 1918. Webb Close recalls Captain Augustus Webb, he died of wounds received when taking part in the famed Charge of the Light Brigade.

Bulmer Road recalls the family who built Brancepeth Castle before it passed to the Nevilles, of whom Ralph Neville gave us Neville Parade. Baliol Road and Baliol Green recall the thirteenth century leader of Scotland John de Balliol. Bruce Road recalls an even better known Scottish king, Robert de Brus (or Bruce) having to cede his lands to the Bishop of Durham on acceding to the throne. Burdon Close recalls the name of Rowland Burdon, the last in a line of lords of the manor of Castle Eden. Cadogan Square is named from George Henry, 5th Earl of Cadogan, a local man who was a trusted advisor to Queen Victoria, Lord Lieutenant of Ireland, and Lord Privy Seal. Castlereagh Close takes the name of Viscount Castlereagh, a man who held a number of government posts. Clarence Chare comes from Clarence Farm and ultimately the Duke of Clarence. Conyers Place recalls Roger de Conyers, hereditary Constable of Durham Castle and Lord of

Bishopton. Crosby Road remembers Brass Crosby, eighteenth century Lord Mayor of London who earned great respect from parliament and first persuaded them to record all future debates and decisions in Hansard.

Former MP and Lord Chancellor John Scott, 1st Earl of Eldon, gave names to both Eldon Close and Eldon Road. Emerson Way recalls William Emerson, the eighteenth century mathematician universally acknowledged as the greatest of his time. Hullock Road recalls Sir John Hullock, a judge and Baron of the Exchequer. Lambton Close honours the work of John George Lambton, whose work in Canada saw him serve as Governor General.

Lyon Walk remembers Eleanor Lyon, who married the 9th Earl of Strathmore and thus becoming an ancestor of the Bowes-Lyon family, and one Elizabeth Bowes-Lyon, the consort of George VI and mother of Queen Elizabeth II. Porter Close recalls Robert Kerr Porter, who married a Russian princess and decorated by the Shah of Persia, although best-known for his vast panoramic paintings of battles. Vane Road recalls Sir Henry Vane who, aged just 24, was Governor of Massachusetts. Westmorland Way recalls Ralph Neville 1st Earl of Westmorland, who built Brancepeth Castle.

No surprise to find collieries to be represented by street names. Buddle Walk recalls John Buddle who, as an owner and manager in the late eighteenth and early nineteenth centuries, did much to increase safety and welfare payments for miners everywhere. While everyone remembers the Davy Lamp, few recall William Reid Clanny, who produced a safety lamp two years before Davy and who had Clanny Road named after him.

Lowery Road recalls Sir Joseph Lowery, knighted for his role in World War I but having a road named after him for his charitable work, particular with miners. Joseph Hopper also did much to aid retiring miners, pioneering the building of homes and having Hopper Road named after him. Joicey Place recalls James Joicey, who purchased a number of collieries in the early twentieth century. Another with a link to miners was Peter Lee, who gave his name to Lee Green and was the first Chairman of a Labour County Council in Britain.

Ramsey Place remembers Thomas Ramsay, a founder of the Durham Miners Association. Westcott Walk is after Brook Foss Westcott, who helped to mediate the end of the Durham miner's strike in 1892. Wilson Walk, after John Wilson, remembers the man who started working in an iron stone quarry for $4^1/_2$d (2p) per day, travelled to America and helped found the National Miners Union.

There are also a smattering of those who fit into no particular category, including former cricketer Robert Bousefield who was also headmaster of the local grammar school and gave his name to Bousefield Crescent. Chapman Walk recalls Abel Chapman, Durham-born author of *Borders and Beyond*, one of several books on birds. Defoe Crescent recalls the author Daniel Defoe, known for the novels *Robinson Crusoe* and *Moll Flanders* but, during his lifetime, better known as a political writer.

Elizabeth Barrett Browning, the poet who married fellow writer Robert Browning, is honoured by Elizabeth Barrett Walk. A semi-legendary woman known as the Saxon Nymph or Elstob, was the first to study the Old English language and was also fluent in nine others and gave her name to Elstob Close. Lilburne Crescent remembers Robert Lilburne, who is best remembers for his role as one of the judges in the trial of Charles I. Lumley Close recalls Lord Lumley, a family who have sent knights to the Crusades, worked alongside several kings and queens, and can trace their ancestry back to early Saxon England.

Particular mention should be made of certain individuals whose achievements simply cannot be summed up in a few words. Bell Walk remembers Gertrude Bell, daughter and granddaughter of men who made their mark in working with iron. For a woman in the nineteenth century to obtain a first-class honours degree from Oxford was laudable enough. Yet she went off to Europe where she succeeded in mountaineering feats which would have made most men quake in their climbing boots. Gertrude then journeyed extensively through Syria, Palestine, and on into the heart of Arabia. For a woman to travel over 500 miles alone in such hostile territory was astonishing and the friendships and contacts she made enabled her to act as liaison during and after the First World War right up to her death in 1926 when she was afforded a burial in Baghdad.

Possibly a less wholesome character, albeit an interesting one, is found behind the name of Barnes Walk. Barnaby Barnes was the son of an Elizabethan bishop of Durham. While his father would have approved in his religious lifestyle, Barnaby also earned some notoriety as something of a swashbuckler, a lover, and a suspected poisoner, while also having an enduring reputation as one of the greatest singers County Durham has ever known.

Colling Walk remembers Charles Colling, a cattle breeder who is less famous than the shorthorn cattle which he helped to make the most widely distributed breed in the world, and they pale in comparison to one particular specimen. The Durham Ox has inspired many a pub name as it travelled the land on a specially built trailer to exhibit this specimen. In 1796 Colling refused an astounding £2,000 for this animal, making much more on his tours. What was so amazing about this creature? Its size. When the poor creature was injured owing to its cart collapsing the carcass weighed over two tons.

Marley Road is after Elsie Marley, proprietor of the Swan public house at Picktree. Maybe her ale was good or bad, we shall never know for there is nothing recorded. She achieved her fame through the rhyme entitled *Elsie Marley*, written about her and telling how she supposedly drowned in a flooded coal pit around 1768.

One pub, just a hundred or so yards from the railway, recalls a comment made by the Sioux chief Sitting Bull when he first saw a railway locomotive. Hence the pub is called the Iron Horse. At the Parsons Centre we find the delightfully-named Turbinia, a pub recalling the work of Sir Charles Parsons, who was knighted for bringing the power and manoeuvrability of the turbine engine to the Royal Navy.

Other public houses here include the Huntsman, a reminder of the days when people wearing pinks could be seen riding their favourite mount following the hounds in pursuit of their chosen prey. In the County we find a pub name inspired by the name of County Durham, the only English county to include the word to differentiate between the city and the county for the vast majority simply add 'shire'.

Newton Bewley

Listed as simply Neuton in 1195, this is a common place name from Old English *niwe tun* which literally means 'the new farmstead'. Correctly this should be seen as 'the newer farmstead', that is one replacing or succeeding another for clearly this has not been 'new' since at least the twelfth century. The second element is almost inevitable with such a common first part. Here the origin is Old French *beau lieu* 'the beautiful place'.

The local name of the Blue House must surely have been down to the colour and yet, as woad was the only blue dye available, this would have been a waste of that precious commodity. Thus it is possible there is an alternative meaning of which we have no clues.

North Shields

Found as Schelles in the fourteenth century, this is from Middle English *schele*, 'a temporary building' that used by a shepherd in summer months. The addition is obvious, for this is opposite South Shields on the banks of the River Tyne.

Norton

Documented as Northtun around the end of the tenth century, this is one of the most common English place names and it is surprising to find no second distinguishing element. Here Old English *north tun* describes 'the northern farmstead' and shows it was named by a settlement south of here.

Locally is Blakeston, which began life as 'the *tun* or farmstead of a man called Bleik'. Harwick comes from Old English *heorde wic*, understood as 'the cattle farm'.

O

Ouston

A name found as Vikilstan in 1244, this features a Scandinavian personal name and Old English *stan* and describes 'the boundary stone of a man called Ulfkell'.

The Brooms is a new pub which appears to have taken an existing name on an old map, the name referring to the tree and not that used to sweep.

Ovington

Domesday records this name as Ulfeton in 1086. Here a Saxon personal name and Old English *ing tun* describes 'the farmstead associated with a man called Wulfa'.

The local name of Nafferton describes 'the farmstead of a man called Nattfari', this is a nickname meaning 'the night traveller'.

P

Pelton

Found as Pelton as early as 1312, this place name features a Saxon personal name and Old English *tun* which describes 'the farmstead of a man called Peopla'.

Locally we find Twizell, a minor name derived from Old English *twisl* or 'fork (of the river)'. Urpeth is found as early as the thirteenth century, it is likely from a personal name and Old English *paeth* and 'the road of a man named Eorl'.

The Bird is a public house is probably named for its image, although a family name cannot be ruled out entirely.

Peterlee

A very recent name created to commemorate the famed trade union leader Peter Lee, who died in 1935.

Crimden Beck House describes 'the stream of the crooked valley'. The only certainty in the name of Castle Eden is the castle gave the addition. Originally this may have been a river name, of Celtic origin meaning 'flowing' or similar. However it could also be a personal name and the river an example of back-formation.

The Royal George is a popular pub name which clearly refers to those four members of the House of Hanover whose reigns lasted from 1660 and the accession of George I until 1830 with the death of George IV. Peterlee's long link with coal mining is marked by the pub named the Five Quarter. This unusual name is taken from the name of the coal seam, the others being the High Main and Yard seams.

The Gamecock takes the name of the birds bred for fighting, not necessarily a place where such took place but an area associated with same.

The Argus Butterfly was named in 1827, the eye-like pattern on its wings designed to scare off possible predators. The name comes from mythology, where the creature called Argus had one hundred eyes, which were transferred to the tail of a peacock following his death. The Shoulder of Mutton most often takes its name from that small feature in the landscape likened to that cut of meat, although there is a chance it was served to travellers.

Hearts of Oak refers to the wood at the heart of the tree which contains no sap, thus it does not need to dry out in the normal way and was highly prized for the building of the ships of the Royal Navy. Hence the term has been used for many years for the British fleet and its men. Having named the wood its most famous creation is seen in the name of *HMS Victory*, Admiral Nelson's flagship. Conversely the Oaklands pub is a very modern name, given to complement the Oakerside estate where it is found.

Piercebridge

The earliest surviving record of this name is as Persebrigc in the middle of the eleventh century. Here the name is derived from Old English *persc brycg* and tells us it was 'the bridge where the osiers grow'.

Locally Carlbury comes from Old English *ceorl bury* and 'the stronghold of the free man', although the first element may also be used as a Saxon personal name. The name is also seen in the Carlbury Arms. Dyance features the Old Scandinavian *dy* which, with the influence of Old French pronunciation, becomes Dyance and describes 'the swampy place'.

Bridge across the River Tees at Piercebridge

Pittington

A name found in the end of the eleventh century as Pittindun. Here Old English *ing dun* follows a Saxon personal name and describes 'the hill associated with a man called Pytta'.

Carlbury Arms, Carlbury

Here is the hamlet of Elemore, a name from Old English describing 'the moor of a man called Ella'. Hallgarth is a combination of Old English *halh* and Old Scandinavian *garthr* telling of 'the enclosure at the nook of land'. Shadforth, recorded as Shadeford in 1190, comes from *sceald ford* 'the shallow ford'. A common Old English stream name, Sherburn is from Old English *scir burna* and describes 'the clear stream'.

Plawsworth

Found as Plauworth in 1297, this name comes from Old English *plaga worth* and telling us it was 'the enclosure for sport or play'. It should be noted some sources suggest the first element may be a personal name.

Nearby Nettlesworth is seen as Nettilsworth, Netyhewrth and Netheworth in the late fourteenth and early fifteenth centuries. These forms are rather late and, while the suffix is clearly *worth*, the personal name is less certain. If we accept these records are correct then this would be Nithbeald or Nidbeald, however there is also a chance this is the name Aetheles which saw the last letter of the previous word *atten* migrate to the following word. Such is also seen in the case of the word 'newt' where 'a newt' was originally written 'an ewt', while the reverse happened with 'orange' where 'a norange' is now written 'an orange'.

There are more pubs known as the Red Lion in England than any other. As with the vast majority of coloured animal names this is heraldic. Most will represent Scotland while the earliest will point to John of Gaunt, the most powerful man in England during the fourteenth century whose descendants fought on opposing sides in the Wars of the Roses for the crown.

Q

Quarrington Hill

A name recorded as Querendune around the end of the twelfth century, this place name comes from Old English *cweorn dun* and describes 'the hill where mill stones were obtained'. The later addition of 'Hill' shows the original meaning was unknown.

Park Hill is found as Pastura del Park in 1342, showing this was an area of enclosed parkland set aside for the grazing of animals, most often for the hunt but that does not seem the case here.

R

Rainton

Record of this name include Raynton, Rayngton, Reinton, Reinuntun, and Rauntone. These early forms show the modern form has been corrupted from the original *inga tun* and a personal name, this being 'the farmstead of the family or followers of a man called Regen'.

Redmarshall

A name found as Reedmershill in 1291, and later as Redmershill, Redmershehille, Redmersell and Redemshall. Here a Saxon personal name and Old English *mersc halh* tells of 'the marsh by the nook of land of a man called Rada'.

Locally we find a common name in Carlton, always describing 'the farmstead of the freemen'. While the name of California is another example of a remoteness name. Recorded as Stillingtune in 1190, Stillington features a personal name with Old English *inga tun* speaking of 'the farmstead of the family or followers of a man called Styfel'.

Redworth

A place name first found in 1183, where it is seen as Redwortha. This comes from Old English *hreod worth* and tells us it was 'the enclosure where reeds grow'.

Romaldkirk

Domesday records this name as Rumoldescherce in 1086, a name from Old Scandinavian *kirja* with a Saxon saint and describing 'the church dedicated to St Rumwald'.

The local pub echoes the theme of the church in the Kirk Inn.

Rookhope

A name which comes from Old English *hroc hop* and describes 'the valley frequented by rooks'. The name is recorded as Rochop in 1242, the local Rookhope Inn also sharing the name.

Ryton-on-Tyne

Ryton is a fairly common name in England, although there are several different meanings. Most often this is Old English *ryge tun* 'the farmstead where rye is grown', although here the existence of the early form as Riton might point to the first element as a personal name such as Riht or Rih.

The local name of Crawcrook united Old English *craw* and Old Scandinavian *krokr* referring to 'the bend in the river frequented by crows'. Here we find Greenside, a name which virtually explains itself as 'the green meadow on the hillside'. Stella is first seen as Stelyngleye, this probably represents Old English *steall leah* 'the stall for cattle in the woodland clearing'. Stella is derived from the Old English *stelling*, a lasting reminder of the farming community for this describes 'the cattle fold'.

Axwell Park is found as Aksheles in 1361, as Axsels in 1382, and as Axsheles in 1396. These all point to an origin of 'the huts of the summer pastures by or made from oak trees'. Barlow features two common elements, *bere hlaw* describing 'the mound or hill near where barley is grown', although it should be realised *bere* is also used as a generic term for grain crops in general, much as everyone refers to the vacuum cleaner as a 'hoover' irrespective of manufacturer.

Chopwell has been recorded since the twelfth century, this is 'the spring or stream of a man called Ceappa'. Crawcrook is probably 'the crook of land frequented by crows', although the name Crow cannot be entirely discounted. Spen is recorded in the modern form as early as 1312. This lack of diversity in the records can often make defining the name difficult and such is the case here. Old English *spind*, which became the dialect words *spine*, *spen*, or *spend*, was used to mean 'turf, greensward'. Although this may indeed be the meaning it would mean this is the only known example of its use in the north of the country.

S

Sacriston

A name recorded as Segrysteynhogh in 1312, this name comes from Middle English *secrestein* and Old English *hoh*. This name tells us it was 'the hill spur of the sacristan', this being the officer charged with the responsibility of the safekeeping of that part of the church, here that being the cathedral of Durham.

Delightful sign welcomes us to Sacriston

Sadberge

The earliest record of this name dates from 1169, exactly as it appears today. This name is Old Scandinavian *set berg* and describes 'the flat topped hill'.

The Buck is a local pub which shows an image of the male deer, probably chosen more for its recognisable image rather than a hunting reference.

St Helen Auckland

Listed as Alclit in 1040, this is a Celtic name referring to 'the rock or cliff on the River Clyde', a river name meaning 'the cleansing one'. However the river here is the Gaunless, which comes from Old Scandinavian *gagnlauss* meaning 'the unprofitable one', that is to say not navigable. The addition tells us the church is dedicated to St Helen.

Regulars at the Bay Horse will see a sign featuring the image of the horse, which can range from light to a mahogany brown and always with a black mane and tail. It may refer to a specific animal, although the image could just as easily have been behind the name.

St Johns Chapel

Clearly this place was named from the local place of worship. What began as a medieval hunting stop, a more permanent and economically viable settlement grew up as a centre for lead mining in the seventeenth century.

Locally we find Irish Hope, *hop* being a 'small valley' and here in the possession of a family from Ireland. Beaufront literally means 'fine brow', a description of the rise giving excellent views to the south across the valley of the Tyne.

The local known as the Blue Bell is a name which shows the close association between the pub and the church. Here the 'bell' is easy to see as the church bell, while blue is representative of Christianity.

Satley

A name recorded as Sateley in 1228, this comes from Old English *set leah* and refers to 'the woodland clearing with a stable or fold'.

Scargill

A place name from Old Scandinavian *skakri gil* and describing 'the ravine frequented by mergansers or similar seabirds'. The name is recorded as Scacreghil in Domesday.

Seaham

Listed as Saeham in the middle of the elveneth century, this name comes from Old English *sae ham* and describes 'the homestead by the sea'.

Street names include Viceroy Street, named after the office of the 6th Marquess of Londonderry, and Stewart Street, the family's name. Similarly Mount Pleasant recalls the grand home in Londonderry. Named because it was only comprised of six properties and thus very small, is Infant Row.

California Row was cut shortly after the California Gold Rush and chosen because it was still very much in the news. Miners from the northeast may well have headed there to make their fortune as indeed they did to the Southern Hemisphere and the promise of a new life in New South Wales, they being remembered by Australia Row. Mining also gave a name to Butcher Row, for an official of the colliery around this time was named Butcher.

Locally we find Dawdon, early records of which are few but which point to a Saxon personal name and Old English *denu* or 'the valley of a man called Dealla'. Slingley, recorded as Slingelawe in 1155 and Slynglawe in 1422, can be defined as 'the *by* or farmstead of a man called Slengi'. Not a personal name in the normal sense but a nickname, this comes from Old Scandinavian *sleng* 'an immature youth, an idler' and also seen in the dialect *sling* 'to go about idling'.

The Kestrel is named after our smallest falcon, commonly seen hovering above the grassy banks alongside our motorways. Another bird, this time a mythological one, is the Phoenix, this name probably represents part of a family crest although which will likely never be known. Noah's Ark is highly unlikely to be religious, the term probably hopes to give the impression the place was suitable for all kinds. The Pemberton Arms is a reminder of the family who once owned the Monkwearmouth Colliery, the place had previously been the Cold Hesledon Inn.

The Inn Between is an obvious play on words, however the question remains what is this establishment 'between'? A glance at the map shows this pub is midway between the harbour and the railway, whether this is mere coincidence is uncertain. The Oddfellows Arms is a name which originated in the Independent Order of Oddfellows, a benevolent society which met at these pubs from the nineteenth century. The Mallard is most often a reference to the famous steam locomotive, however in Seaham the image and the name refer to the common waterfowl.

Clearly a military reference, the Volunteer Arms could refer to virtually any period of conflict in British history. Today there is no sign hanging depicting a uniform. Although most of this name depict those regiments from the eighteenth century who were to defend our shores against the threat of invasion by Napoleon, it could also refer to the Volunteer Training Corps of World War I or the Local Defence Volunteers (also known as the Home Guard) of World War II. What some would consider an overly familiar reference to our nation's longest reigning monarch is the Queen Vic.

Sedgefield

Listed as Ceddesfeld in the middle of the eleventh century, this name comes from a Saxon personal name and Old English *feld* and telling us this was 'the open land of a man called Secci'.

Here is the hamlet of Bradbury, a common place name which, as here, is invaribly from Old English *brad burh* and 'the broad fortified place'. Mordon is recorded as Morton in the thirteenth century, showing this is 'the farmstead on or by the moorland'. The Isle is an area formed by the confluence of the rivers Skerne and Rushyford. Butterwick describes 'the dairy farm with good butter'. What was 'the old arable land' is now known as Old Acres. Ten O'Clock Barn may seem an unusual name, however it is easy to see the origin when it is pointed out it refers to its direction.

Embleton is from *elim denu* or 'the valley of the elm trees'. High Swainston, Middle Swainston and Low Swainston share a name meaning 'the *tun* or farmstead of a man called Swein'. Bedlam Gill has a derogatory name for a valley, similarly 'the ravine of the shoemaker' is today marked as Cobbler's Gill. Swallow Hole tells us it was where a stream disappears undergrown.

Fishburn comes from *fisc burna* or 'the stream where fish are caught'. Harap Hill comes from *hara hop* or 'the valley frequented by hares'. Foxton also comes from a mammal, no surprise to find 'the valley frequented by foxes'. From *scrot tun* comes Shotton or 'the farmstead at the steep hill'. Hardwick tells us it was 'the dairy farm'. From *lad tun* or 'the farmstead on the stream' comes the names of East Layton, West Layton, and South Layton.

Bono Retiro is Italian referring to 'the Gothic retreat' in the grounds of Hardwick Hall. Dial Hill Plantation had a huge sundial carved into the turf. Lizards is from *laesow* meaning 'meadow, pastureland'. Salters Lane is ancient route from the Tees to Trimdon. Tinkler's Moor Plantation show where a group of itinerant tinkers once camped.

Mordon comes from *mor dun* 'the swamp by the hill'. Oldacres almost describes exactly what it is, only the meaning of *acres* has changed to give 'the old cultivated land'. Ryal comes from *ryge halh* 'the nook of land where rye is grown'. Herdewyk on Sea describes 'the specialised farm of the herdsman near the sea', note this refers to one who tends to the flock hence sheep not cattle.

Shadforth

The earliest record of this name dates from 1183 as Shaldeford. This comes from Old English *sceald ford* and describes 'the shallow ford'.

Sheraton

Found in a document from around 1040 as Scurufatun, this comes from Old English *tun* and a Scandinavian personal name and telling us of 'the farmstead of a man called Skurfa'.

Sherburn

A fairly common place name which comes from Old English *scir burna* and describes 'the place at the bright or clear stream'. This name is found as Scireburne around 1170.

Shildon

Found as Sciluedon in a document dating from 1214, this name comes from Old English *scylfe dun* and describes 'the shelved hill'.

Shildon styles itself 'Cradle of the Railways'

Locally we find Brusselton, a name which has no early records making definition rather difficult. However by association with similar names throughout England we can see a number of potential beginnings which rely a lot of whether the 'l' ever existed originally, it may also be a remnant of *leah* or simply be an error. Thus this could be Beorht's *leah tun* 'the farmstead in the woodland clearing of a man called Beorht' or possibly 'the farmstead of a man called Burhwulf'.

Shildon's millennium gate

The Fox and Hounds is instantly associated with hunting, however that it exists outside areas associated with hunting indicates it was more likely chosen for making an attractive sign. It is always a good selling point to show support for the monarchy for it also shows patriotism and the Queens Head does just that in the form of Queen Adelaide, consort of William IV who is also named by Adelaide Terrace.

The Locomotive is a general reference to the railways, while the Timothy Hackworth remembers the former blacksmith who became involved in locomotive production from 1808. The Jubilee Inn is in the region of Jubilee Avenue and Coronation Avenue, the pub continuing the theme of the roads.

Shincliffe

A name recorded as Scinneclif around the end of the eleventh century. This comes from Old English *scinna clif* and rather intriguingly speaks of 'the cliff haunted by a phantom or demon'.

The local name of Whitwell is easy to see as coming from Old English *hwit wella* and speaking of 'the white spring or stream'.

As the signwriter was obviously aware roses do not grow on trees. Thus the sign at the Rose Tree Inn shows the flower of a single rose in the foreground with a large beech tree in the distance. Here the tree would have been a landmark while the rose most likely represents England.

The Seven Stars on the High Street is an interesting name. Three of the four pub signs do not hang out from the building but are fastened flush to the wall. These three are identical and in the bottom left hand corner of each is the unmistakable image of seven stars forming the constellation of Ursa Major, also referred to as the Great Bear, the Big Dipper, and the Plough. The latter is very interesting because the fourth sign is very different, for not only does it hang away from the building but also features a very different image. Here we see the silhouette of the oxen and the ploughman turning over the soil. Together they tell us the original name of the pub was the Plough.

Shotton

Recorded as Sotton around 1165, this name comes from Old English *sceot tun* and tells us of 'the farmstead on or near a steep slope'.

Silksworth

Listed as Sylceswurthe in the middle of the eleventh century and as Sulkeswrthe at the end of the twelfth century, this name comes from Old English *worth* and a Saxon personal name and referring to 'the enclosure of a man called Sigelac'.

To the north is the later development of New Silksworth. Here street names were planned when, towards the end of the nineteenth century, housing was provided for pit workers. Five ladies of the Londonderry family gave their name to five streets: Cornelia Street, Mary Street, Aline Street, Frances Street, and Maria Street.

Skerne (River)

Found as Schyrna at the end of the twelfth century, this name comes from Old Scandinavian *skirn* or Old English *scir*, describing 'the clear or bright stream'. The Saxon *scir* seems more likely, with the pronunciation influenced by the Scandinavians.

Sleightholme

A name found as Slethholm in 1234, this comes from Old Scandinavian *slettr holme* and describes 'the level raised ground'.

Sockburn

Documented as Soccabyrig in 780, as Socceburg in 1050, and as Sockburn in 1130, here is a name from Old English *burg* following a Saxon personal name and telling of 'the fortified place of a man called Socca'.

The local name of Girsby is of Old Scandinavian origin, *griss by* describing 'the farmstead known for the rearing of young pigs'.

South Shields

Seen as Schelles in 1364, this comes from Middle English *schele*, 'a temporary building' that is one used in summer, most likely by a shepherd. The addition is obvious, it lies to the south of the River Tyne and opposite North Shields on the other bank of the river.

Locally we find the name of Harton, this comes from *heort tun* or 'the farmstead where harts or stags are seen'. Simonside, recorded as

Symonseth in 1313, 'the slope associated with a man called Sigemund' – the place was originally known as Preston 'the farmstead of the priests'.

Wrekendike is found as Vrakendic in 1150 and as Wrakendyk in 1225, a name from Old English *wraeccna dic* or 'the dyke of the fugitives'. Found along the line of a former Roman road, the ditch or dike would probably have been linked to the Roman occupation, the fugitives would have been the later Saxon or Scandinavian. Swainston is a combination of a Scandinavian personal name and Old English *tun* and tells of 'the farmstead of a man called Swegen'.

Spennymoor

Irrespective whether the name is Old English or Old Scandinavian the origin is the same. From *spenning mor* this name describes 'the moor with a fence' and appears as Spendingmor around 1336.

Street names begin with Jackson Street, a name which can be traced

Spennymoor's town hall building

to 1603. Documents from this time show some 170 acres of land granted by Thomas and John Jackson. Another land-owner is remembered by Shafto Street. The buyer one Mark Shafto the third son of Robert Shafto, Sheriff of Newcastle.

Locally we find Hooker Gate, a name seen in 1587 as Hucker-gaite and as Hookegate in 1596. Here is a personal name and Old English *gata* describing 'the way of a man called Huckster'. Penny Gil, Binchester and Hillingdon are three pub names which have taken nearby place names. While it is hardly imaginative it does make sense for prospective customers know exactly where they could be found in times

The Penny Gill public house, Spennymoor

when addresses and postcodes were neither necessary nor known.

The Pit Laddie is a pub name which remembers the boys who worked in the pits in former times. When it was opened the Bridge Inn was probably in sight of the bridge across the river. The Shafto Inn appears to suggest it is named from the nursery rhyme which begins "Bobby Shafto's gone to sea, silver buckles on his knee", but actually has the same origin as Shafto Street. The Railway is not only self-explanatory but also datable for the pub cannot have existed before the railway arrived here. The same is true of the North Eastern, for this refers to the local line.

A horse racing link is seen in the Winning Post and the Voltiguer, the latter the name of the horse which won the Epsom Derby in 1850. Architect Anthony Salvin (1799-1881), an expert on medieval buildings, of castles and country houses, is commemorated by the Salvin Arms. The Ash Tree is one of the many pubs which use a prominent nearby tree as a name to act as a marker. The image of the attractive bird was undoubtedly the reason for the naming of the Kingfisher.

Powerful landowners, the dukes of Newcastle are known for the greyhound which is seen in their coat of arms, this is the basis for the name of the Greyhound Inn. Alliteration is a marketing tool used by advertisers the world over, however this is not a new idea but has been used by inn-keepers for centuries. Furthermore to unite two seemingly unrelated items is also seen as being an obvious pub name. Both criteria are seen in the name of the Frog and Ferret.

Staindrop

Listed as Standropa around 1040, this name comes from Old English *staener hop* and describes 'the valley with stony ground'.

Ingleton is a minor name whose early records are little different to the modern form. It is thought to represent a Saxon settlement in a region where Scandinavians predominated, thus 'the farmstead of the Angles'. Keverston's earliest record is as Kevyrston, a shortened version of 'the *tun* or farmstead of Ceolfrith'. Little Newsham derives its name from 'the new houses', the addition distinguishes this from a similarly named place in neighbouring Yorkshire.

Raby is an Old Scandinavian name describing 'the farmstead near the landmark'. The most obvious feature at Raby is the castle. In a 200 hundred acre park it was built by John Neville, 3rd Baron Neville de Raby, between 1367 and 1390. Cecily Neville, mother of Edward IV and Richard III, was born here. Sir Henry Vane bought this castle in 1626 and extensive alterations followed. Today it has Grade I listing and is known for its size and for the splendid collection of art found within.

The earliest known record of Snotterton is from 1411, where the name is exactly as it appears today, means this can only be defined as 'the *tun* or farmstead of a man called Snotter'. Keverstone is Kevrestone in 1306 and Kewreston in 1330, this is difficult to define but possibly comes from Old English '*tun* or farmstead of a man called Cenfrith'. Here the minor name of Shotton is from Old English '*tun* or farmstead of a man called Scott'. Whether the record is as Gibbs Knees or Gibbs Neese this can be defined as 'the headland of a man called Iba'.

The Wheatsheaf Inn is one of the many pubs where the name comes from a coat of arms. Here the device is representative of either the Worshipful Company of Bakers or of the Brewers' Company. Another heraldic image is that of the Black Swan, dating from a time when it was believed all swans were white it came to mean one of rare or unique talents. After Captain James Cook discovered Australia and its indigenous black swans this meaning was forgotten, indeed more recent creations probably refer to some link to the southern continent.

Stainton

Old English *stan tun* gave this place a name recorded as Staynton around the middle of the twelfth century. This name speaks of the farmstead on stony ground and would be Stanton had the first element not been influenced by Old Scandinavian *steinn*.

Stainton (Great & Little)

As with the preceding entry, here Old Scandinavian *steinn* has influenced an Old English name from *staning tun* and refers to 'the farmstead at the stony place'. This place name was recorded as Staninctona in 1091, the later additions are self-explanatory.

At Little Stainton is found Nova Scotia Plantation, a name transferred from that part of distant Canada known as Nova Scotia or 'New Scotland', a reminder of who settled here. However this is not to say there were any more Scotsmen here than elsewhere, for this is a remoteness name. A look at any map will show parishes are never symmetrical, indeed many resemble the shapeless blob of protoplasm of the amoeba drawn in biology lessons of our childhood. Such shapes invariably mean there is at least one arm of the parish which is further removed from the central hub than others and this was often given such a 'remoteness' name by those who had to walk this comparatively great distance to work.

Streatlam is almost certainly a 'the woodland clearing on or by the Roman road'. It is found as Streatlea in the thirteenth century and would be from *straet leah*.

Stanhope

A name which comes from Old English *stan hop* and which describes the 'valley of stony ground'. The name is recorded as Stanhopa in 1183.

Ferryfield means exactly what it seems, the 'open land where the ferry crosses the river (Wear)'. Frosterley is still easily seen as 'the *leah* or woodland clearing associated with the forester'. Bewdley is from Old French *beau* and Old English *leah* meaning 'the beautiful woodland clearing', where the pronunciation of the 'd' can only have come from an error in spelling.

Bolts Law puts together *botl* and *hlaw* telling us to look out for 'the hill or mound with a building'. With a suffix from Old English *leah*,

Brotherlee describes 'the woodland clearing of a Brother', not a religious character or a relative but a fairly common Scandinavian personal name. Many towns and villages have a Pinfold Street, yet not often is this found as a minor place name. Old English *pin fald* describes 'the fold of the pinner', the man charged with rounding up and holding stray livestock awaiting collection and payment of the fine by the owner.

Ravensfield speaks for itself as 'the open land frequented by ravens'. The earliest record of Snowhope is as Snawhopkerr in 1382, this being 'the marsh by the valley where the snow lies'. While the reference to the marsh has now been dropped, the sheltered valley still shades the winter sun and any lying snow still lingers.

Rookhope has changed little since the earliest known record from 1190 as Rokehope, this being 'the small valley frequented by rooks'. Slatyford is found as Slaterforth in 1382, the only early record of note and one describing 'the ford of a man called Slater', here the name is neither a family name nor a reference to his occupation. Spain's Field is found as Spaynesfeld in 1382, the name dating from around two centuries earlier when the manor is recorded in Domesday as a possession of Henry de Ispania. Stotfield Burn comes from Old English *stod feld burna* 'the stream by the open land where horses are reared'. Westernhope reminds is this was 'the valley where whetstones are found'.

Catterick Moss is seen as Katericksaltere in 1331 and Catryk in 1382, a name from a Celtic tongue describing 'the cataract or waterfall', the addition of 'moss' is very recent and distinguishes this from Catterick in Yorkshire. Eastgate is, as the name still very much suggests, 'the eastern way or route'. Hanging Wells does indeed refer to 'a hanging spring', telling us it reaches the surface at an overhanging rock. Waskerley Beck has a name referring to 'the woodland clearing at the wet marsh'.

Harelaw combines two Old English elements, where *har hlaw* describe 'the tumulus or mound with a stone marker'. Harrowbank is a difficult name to define, however two possibilities stand out: either this is *hearge* and 'the bank of land at the heathen grove' or *harewe* 'the bank of land worked with a harrow'. Rogerley is recorded as Rogerleia in the fourteenth century, this being 'the woodland clearing of a man called Roger'.

The Grey Bull is a public house telling of the importance of farming to the area. While most coloured animals are heraldic, here the name would point to the rearing of cattle, while the colour is probably derived from the sign rather than any particular animal. The Bonny Moorhen has little to do with the waterfowl, this is a song reflecting the Jacobite period of Scottish history. It tells of the fugitive prince pursued by the red coat troops, his identity revealed in the lines "And she's a' fine colours, but nane o' them blue, She's red an' she's white, an' she's green an' she's grey", the tartan of the Stuarts.

Stanley

A very common place name which always comes from Old English *stan leah* and refers to 'the stony woodland clearing'. The name is found in a document dated 1130 as Stanlega.

Hare Law is a minor name originally seen as Herilaw, Harelaw, and Harlaw, and is most likely from Old English *har hlaw* and describing 'the mound or hill at a boundary'. Found as Stellyngton in the thirteenth century, this comes from Old English *steall ing tun* or 'the farmstead of the cattle stall'.

Bishopley tells us it was a possession of a religious order, specifically 'the woodland clearing of the bishop'. Quaking Houses is a reminder of the Quaker House Pit, the cottages here housed the miners and their families. The pit itself took the name of the nearby Quaking House Farm.

Pub names begin with Hill Top, itself taken from a local place name of obvious meaning. The Crown and Thistle shows the proximity to the border with Scotland, the Crown representing England and the Thistle Scotland and the union of the two countries with the accession of James I of England and VI of Scotland. The close proximity of the border is marked by the name of the Highlander.

Very few pubs named the Bluebell Inn are actually a reference to the spring flower. The true origin is religious and the name two words, with the bell representing the church and the colour Christianity. First seen as an heraldic reference to Richard III, the Blue Boar has been seen as a pub name since the fifteenth century. The Townley family can trace their ancestry back to the Norman Conquest, they have given their name to the Townley Arms. The Earl Grey Inn is named after Charles Grey, 2nd Earl Grey who served as prime minister (1830-4) during which time he helped with the abolition of the slave trade.

Both the Highwayman and the Smiths Arms are self-explanatory, unlike the Plainsman which must be one of the very few pubs in the land named after a Hollywood film. Released in 1937, *The Plainsman* starred Gary Cooper as Wild Bill Hickok.The Ox Inn is named after the Durham Ox, bred by Charles Collinge of Ketton by the age of six it weighed almost two tons. Collinge bought the beast for the (then) huge some of £250 but still turned a massive profit by touring the country with the beast in a specially constructed cart.

Carts from Tanfield Colliery would cross the nearby Causey Arch. Built in 1726, this single-span bridge was the longest in the country at the time and today is the oldest surviving single-arch bridge in the world. It carried two tracks, the main way carrying full trucks to the River Tyne and the bye way allowing the returning empties back to Tanfield. The bridge was built by stonemason Ralph Wood at a cost of £12,000 and funded by a group of local coal-owners known as the Grand Allies. Ironically wear and tear on the bridge was reduced by the closure of Tanfield Colliery when it was destroyed by fire twelve years after the bridge was opened to traffic.

Startforth

Listed as Stretford in the middle of the eleventh century, this name comes from Old English *straet ford* and speaks of 'the ford on the Roman road'.

Stockton-on-Tees

Seen as Stocton in 1996, this is a name found in several places and always from *stoc tun* and here best understood as 'the farmstead of the outlying place'. The addition of the river name, discussed under its own entry, shows how common this name is.

Preston-on-Tees also adds the name of the river to a common place name, here Old English *preost tun* speaks of 'the farmstead of the priests'. The minor name of Hartburn is a common one, most of which are found in former hunting lands, this comes from *heort burna* and describes 'the stream frequented by stags'. East Hartburn, almost certainly a name transferred here when a section of the community relocated to not only pastures new but new land for growing crops. Finkel Street takes its name from a dialect word describing an area where 'fennel' grows. Hornleys is easy to see as 'the horn-shaped meadows', while Ropner Park was an area of land donated to the town in 1893 by Sir Robert Ropner.

Redmarshall comes from *hreod meres hyll* 'the reedy marsh by the hill'. Holstone House was constructed on a place already named for 'the hollow stone' and undoubtedly a boundary marker.

Stranton

Recorded as the modern form as early as 1159, Stranton describes 'the farmstead on the shore'.

Here we find Brierton, meaning 'the briar farmstead'; Oughton began as 'the *tun* or farmstead of a man called Ofa'; Tunstall is from *tun steall* 'the farmstead with a stall'; and The Slake tells us it was 'muddy ground, tidal flats'. Clearly West Hartlepool is a second place describing 'the pool or bay near the peninsula frequented by stags', the addition for distinction. A leading light of West Hartlepool was Ralph Ward Jackson, whose name was taken for Jackson's Lodge, Jackson Dock and Ward Jackson Park. Seaton Carew is from *sae tun*, an accurate description of 'the farmstead by the sea' which was associated with Robert de Carew in the early twelfth century. North Gare Breakwater features the dialect element *gair* or 'a triangular piece of land'.

Sunderland

One of the newest cities in the land has a name dating from Saxon times. Here Old English *sundor land* describes 'the detached estate'. Clearly the name has changed by just one letter over a period well in excess of a thousand years, with the earliest record of the modern form as early as 1168.

Street names tell their own version of local history and have become increasingly important as a tool to aid the work of historians, archivists, archaeologists, etc. In Sunderland we find the usual collection of street names and also those which are lasting reminders of those who have contributed to the history.

We start with Abbs Street, an unusual surname which may could refer to George Cooper Abbs who owned a large house nearby, or brewery owner Stafford Abbs. Aiskell Street takes the name of Edward Aiskell, an eighteenth century coal agent. Alder Street takes the name of developer Stanley Alder. Andersons Passage probably takes the name of a woman who kept a shop selling oranges and apples.

Ann Street was cut on land owned by General Aylmer and probably named after his wife Anne Harrison despite the spelling error. Similarly Aylmer Street took the name of the general. Arras Lane remembers John Arras, who held the land around here. Arrol Park recalls the builders of the Queen Alexandra Bridge, Sir William Arrol & Co Ltd. Ashburne Court remembers Ashburne House, the former home of the bankers the Backhouse family. Ayres Quay and Ayres Quay Road take the name of the seventeenth century developer Robert Ayre.

Banes Lane is named after John Bayne (also given as Bain), a coal fitter who rented property here in the first part of the eighteenth century. Balfour Street recalls David Balfour, civil engineer for the local district highway board and agent for the Bax-Ironside estate. Balmer Street took the name of W.P & M.E. Balmer, a local building firm. Barclay Street takes the name of Maria Williamson's husband, her family were landowners for many generations. Either former MP Beatrice Webb or a member of the Williamson family gave her name to Beatrice Street.

Betty Cash's Lane recalls Mrs Elizabeth Cash, resident here in the eighteenth century. Birds Lane remembers Richard Bird, who had the ropery in Ropery Lane. Bowes Quay was built by Ralph Bowes and was working in May 1801. Bonnersfield is alongside the river, a name which recalls the quayside raff yard (timber yard) of a Mr Bonner of Southwick. Brandling Street recalls the name of the family who helped bring the railways here from Newcastle and South Shields.

Bright Street is named after John Bright, a prominent member of the cabinet under prime minister William Gladstone. Viscount Edward Cardwell also served in that cabinet, hence the name of Cardwell Street. Cannys Place and Cannys Square commemorate the life of Captain John Canney, a marine stores dealer who always wore a black patch on his lip to cover an old injury which resulted in him being known as 'Patchy' Canney.

Burdon Lane remembers Thomas Burdon, who bought a house around the corner in Church Street in 1820. Burleigh Street is named after the eighteenth century Quaker, coal fitter, and wine merchant Mark Burleigh. Burlinson Street is named after Mr Burlinson the baker who lived here at the end of the nineteenth century. Burnaby Street took the name of Frederick Gustavus Burnaby, who joined the Horse Guards in 1842 aged 16 and who wrote extensively about his military travels and adventures for the next forty years.

Byleth Lane can only be named after a Mr Byleth, recorded as living in Love Lane in a document dated 1739. Cashs Lane took the name of the Mr Cash who owned the clothes shop on the corner of the lane. Chipchase Street may recall landowner Dr Chipchase Grey, although Chipchase is also a surname found several times in the city's history. Clanny Street was named after Dr William Clanny who had a practice here and who would have been remembered for inventing a miner's lamp had not Sir Humphrey Davy produced a better working example.

Cooper Street was named after Chartist MP Thomas Cooper. Cornforths Lane takes the name of John Cornforth, a ship owner and landowner. Cromwell Street was not named after Oliver Cromwell but, as it was originally known as Robert Cromwell Street, was named after either his father or his son who were both named Robert. Local councillors gave their names with Martha Orr and Orr Avenue, John Norman and Norman Avenue, George Park and Park Avenue, Joseph Davision and Davison Avenue, James Higgins and Higgins Avenue, William Emmerson and Emmerson Avenue, George Robinson and Robinson Avenue, and George Byrne and Byrne Avenue.

Derby Street is named after Edward, 14th Earl of Derby who served three short terms as prime minister. Dixon Square can be traced back to Thomas Dixon, a shipbuilder who was recorded here in 1781. Drysdales Entry takes the name of William Drysdale, a cooper who rented property near here in the late eighteenth century. Dunns Passage dates from the middle of the eighteenth century, when William Dunn had a pawnbroker's shop here. Earnwill Avenue was named after Councillor Earnest William Thompson.

Elizabeth Street was named by the builder, Stanley Alder, after his wife Elizabeth. Similarly Fox Street was named after a builder by the name of Fox, who also used the name of his daughter for Evelyn Street. Elwin Terrace commemorates Alderman Robert Elwin, who served as a local councillor for twenty years of the nineteenth century. Ewing Street recalls Professor Sir James Ewing, Professor of Engineering at Dundee University, and later Professor of Mechanism and Applied Mathematics at Cambridge University.

Not only people but buildings offered suggestions for a name. Athenaeum Street came from the building housing the library and museum. Barrack Street was known as such from the early nineteenth century, the barracks being found opposite. Beehive Lane was named after the Beehive Inn which once stood here. Cage Hill is named after a lock up premises which in effect was little more than an iron cage.

Themes for new streets are commonplace today, however this is not a new idea. Indeed while the twenty-first century seems unable to stray far from a list of flowers, trees, birds, writers, etc., early generations were far more imaginative. For example running south off Chester Road we find seven streets all named after English market towns. However look closely and they also follow the ABC pattern from east to west. While alphabetical order is seen regularly, to find one street for each letter is quite rare as in Abingdon Street, Barnard Street, Colchester Terrace, Dunbar Street, Eastfield Street, Farnham Terrace, and Guisborough Street.

The nobility were once a favourite subject, hence we find Albert Street after the consort of Queen Victoria; Adelaide Street, the consort of William IV; Leopold Street, son of Queen Victoria Duke of Albany and Earl of Clarence; Alfred Street, son of Queen Victoria and Duke of Saxe-Coburg and Gotha; Alexandra Road remembers the Danish princess who became the consort of Edward VII; Alice Street, Princess Alice, daughter of Queen Victoria; Argyle Square and Argyle Street take the name of the 9th Duke of Argyll, husband of Princess Louise, daughter of Queen Victoria; the Duke of Argyll was also Marques of Lorne, hence the name of Lorne Terrace.

Avondale Terrace recalls Prince Albert, Duke of Clarence and Avondale; Bloomfield Street remembers the daughter of the 7th Baron Georgina, known as Baroness Bloomfield; Brougham Street recalls Henry Brougham, first Baron Brougham and Vaux (1778-1868) a campaigner against slavery. Broughs Lane was cut on land owned by William Brough until 1691. Edward Street takes the name of Edward VII.

The Londonderry family gave their name to a number of streets including Lord Street, Londonderry Street, Cornelia Street, Aline Street, Maria Street, Frances Street, and Mary Street. Dame Dorothy Street was named after local benefactor Lady Dorothy Williamson, who left money to be invested for several charities supporting the poor. Douro Terrace takes the name of the Duke of Wellington, who was also Marquis of Douro.

Duncan Street was after Adam, Viscount Duncan remembered as a hero when a Royal Navy officer who earned a great victory against the Dutch. Dundas Street remembers Maurice Fitzhardinge, First Sea Lord Dundas who provided the funds for the RNLI's first lifeboat. Ettrick Grove is after Walter Ettrick, who was appointed Collector of Customs here in the middle of the seventeenth century. He accumulated sufficient wealth to buy many properties include quays, houses and parts of the Barnes estate. Land owned by William Ettrick was developed and named Ettirck Place, Ettrick Square, and Ettrick Quay.

Religion also figures prominently and for a variety of reasons. All Saints Court is on land owned by All Saints Church. Barrington Street recalls the Shute Barrington, former Bishop of Durham. Benedict Street can be traced back to Benedict Biscop, a nobleman whose religious conversion saw him building monasteries at Monkwearmouth and Jarrow. Bernard Street commemorates the work of Bernard Gilpin, rector of St Michaels Church in the sixteenth century. Bishopton Street is on land once owned by Dr Tatham, whose son Ralph became rector of Bishopton. Bowlby Street takes the name of the Reverend Thomas Bowlby. Eden Street takes the name of the rector of St Michaels, the Rev J P Eden.

There are those which commemorate events, some better known than others. One of the lesser known gave a name to Amethyst Street. During the Communist uprising of 1949 the British frigate *HMS Amethyst* was in port at Shanghai. With Mao Tse-Tung's men advancing they sailed for Nanking, however the passage was fraught with danger and 43 men were killed under a hail of bullets, including the captain and ship's doctor, the vessel running aground. Many attempts were made to rescue the survivors until, after four months of misery, the ship made a break for freedom under cover of darkness. Their perils were not at an end the craft shelled a number of times during the 140-mile voyage to Hong Kong, yet somehow they gained the security of the British holding without incurring further losses.

Cawnpore Square took the name of the Indian city of Kanpur where, in 1857, the Bengal Army sepoys rose up against the British Army and killed all the men and imprisoned the women and children. Concorde Square was laid out in 1967, when the world's first supersonic jet airliner, the Anglo-French Concorde, made the newspapers almost daily. Coronation Street was named such in 1821 when King George IV was crowned and every window in the town was illuminated that night with a candle or oil lamp.

There are also those names which fit into no particular category. Barleycorn Place takes the name from the measurement equal to half an inch, although it is a very short street it does measure a good deal longer than a barleycorn. Tradition has it Beggars Bank was where those unfortunates scrambled down to cross out over the boundary from one judiciary to another, thus evading capture.

Blue Anchor Yard took its name from the Blue Anchor Inn, although this is only an unofficial name for it is correctly referred to as the Custom House Yard, again named after a building, this time the

Custom House. Bodlewell Lane is named after the spring which supplied water of excellent quality, indeed it was so good a price was put on it. Hence for each skeel (two gallons) customers had to pay one bodle (half a farthing).

Crowtree Lane was named after the trees which were found alongside the lane, themselves marked out by the birds (more likely rooks than crows) which frequented them. Essen Way takes the name of Essen in Germany, Sunderland is twinned with this place and also the French town of Saint-Nazaire.

The minor name of Coxgreen features a Middle English suffix *grene* which can be understood as 'village green', with a first element having a number of possibilities. Fulwell is a common name found throughout the land and from *ful wella* 'the muddy or dirty spring or stream'. Herrington features a Saxon personal name and Old English *ing tun* to speak of 'the farmstead of the family or followers of a man called Heor'. Offerton is easily seen as 'the upper farmstead' from *ofer tun*. The district of Pallion is found as Pavylloun in the fourteenth century, however it is the record from 1453 as Pavilyon which is clearly from Old French *pavillon* a building erected for special purposes, which must have been a retreat for the monks.

Roker is a suburb of this city, best known to the nation as the old home of the local football club at Roker Park. Early records are unavailable, however it is still possible to define it if this has followed the normal evolution of names in the area. This probably represents Old English *hroc* and Old Scandinavian *kjarr* or 'the marsh frequented by rooks', although the first element could also represent a personal name.

Ryhope is south of the city, where Old English *hop* follows a Saxon personal name and tells of 'the small valley of a man called Raf'. Howick is from Old Scandinavian *har vik* 'the high inlet'.

Sunniside

Just three miles southwest of Gateshead, early forms of this name begin with Sonnyngside and Sunnysyde in the fourteenth century. Not a 'pleasant' name, the first element is a Saxon personal name and describes 'the hillside of a man called Sunn or Sunna'.

T

Tanfield

Listed as Tamrfeld in 1179, this name describes 'the open land on the River Team'. Here the Celtic river name, discussed under its own entry, is followed by Old English *feld*. To the north is Tantobie, which may well feature the same first element, although this would have been transferred, with the suffix from Old Scandinavian *by* or 'farmstead'.

Minor names here include Foulbridge House, where neither the house nor the bridge was 'foul' but the *ful* or 'muddy' stream which ran here. A simpler origin is found in Andrews House, named after the family who built it. Beckley, recorded as Bekkeley in 1344, unites a Saxon personal name and Old English *leah* in 'the woodland clearing of a man called Becca'. Causey Hall is an Old Scandinavian name speaking of 'the cold or exposed farm', the modern form clearly influenced by *cawse* 'causeway'.

Lintz Ford is a combination of *hlinc* and *ford*, Old English for 'the ford by the ridge of land'.

Regulars enjoying a glass of their favourite tipple at the Peacock, may like to know the name has at least four possible origins. Clearly for sign-painters this provides an attractive image and, for the same reason, was chosen to symbolise the Manners family, dukes of Rutland. It could also represent a surname, although none are known locally, or even a nickname for someone overly proud of their appearance. The Packhorse reminds us of the role played by inns in the days when they served as staging posts for goods in transit.

Team (River)

This is an ancient Celtic name, thought to speak of 'the dark one'. Here the river name is likened to similar names across the land including Thames, Teme, Tame, Thame, and Tamar.

Tees (River)

A Celtic or earlier name which speaks of 'the surging river'.

River Tees at Piercebridge

Thornley

Listed as Thornhlawa in 1071, this name comes from Old English *thorn hlaw* and describes 'the thorn tree hill or mound'.

Wheatley is a simple enough name, derived from Old English *hwaet leah* and telling of 'the woodland clearing where wheat is grown'.

Thorpe Thewles

Old English *thorp* describes an 'outlying farmstead', a part of the original settlement which grew up away from the main centre of population. In most cases this would have begun purely for agriculture, initially seasonal, eventually becoming permanent. The addition is unusual for it does not seem to be manorial. However it is a Middle English word, *thewles* meaning 'vicious, immoral'. How this translates to an addition is unclear, indeed we shall likely never know.

Here we find Grindon, a minor place name with an early record of Grendon confirming initial suspicions that this is from *grene dun* and describing 'the green hill'. Fulthorpe is from *ful thorp* 'the foul or dirty outlying farmstead'. Wynyard Hall took a name which was here well before the hall was built, the name coming from *win geard* and describing 'the vineyard'. Built in the nineteenth century, Lion Bridge allowed travellers access to Wynyard Hall. It was named for the five lions found on the coat of arms of the owners of the hall, the Marquesses of Londonderry. Whitton is a Saxon place name, telling us it was 'the farmstead of a man called Hwita' or, less likely, 'the white farmstead' from *hwit tun*. The field name of Dog Kennel always tells us it was big enough for a pack of hounds.

Tow Law

Recorded as Tollawe in 1423, this name comes from Old English *tot hlaw* and refers to 'the look out hill or mound'.

The New Market public house speaks for itself, although the market has not been 'new' for many years. While the railway no longer runs here the Old Station Inn has found a new life as licensed premises.

Trimdon

Found as Tremeldon in 1196, this name probably comes from Old English *treow mael dun* and describes 'the hill with a wooden cross'.

The local is the Bird in Hand, a welcoming name reminding us be thankful for what we have more than what we aspire to.

Tudhoe

The earliest surviving record of this name dates from 1279 when the name appears exactly as it does today. Here is a Saxon personal name and Old English *hoh*, together referring to 'the hill spur of a man called Tuda'.

Tweed, River

A river name documented as Tuidi fluminis in 730 by the Venerable Bede, with later records of Tuidon in 890, Twiode at the end of the tenth century, Tweoda around 1050, and Tweda in 1108. An important river since it began flowing again following the retreat of the ice sheet some ten thousand years ago, this is reflected in its name which can be traced back to this time through its name. Here the theoretical Proto-Indo-European language, the mother of the many tongues spoken, as the name suggests, across the European and Indian continental land masses, is the key. While that language is unknown, we can look at those ancient languages which are known, such as Latin, Greek, the Celtic group and Sanskrit. Most river names are highly simplistic and here the root is related to Sanskrit *tavas* meaning 'powerful', an apt description of the Tweed, particularly its upper reaches.

Tyne, River

As with the previous entry, another Celtic or pre-Celtic river name which can be traced back to the earliest development of the Indo-European languages. Surviving records of this name include Tina from the second century, Tinus in 730, Tine in 875, and Tinam Australem in 1130 – this last form simply means 'the southern Tyne, distinguishing it from that found in Scotland. Old English *thinan* meant 'dissolve', itself from the root *ti* with the same meaning and related to an unknown root which undoubtedly meant simply 'river' and telling this was flowing water.

U

Uldale

Listed as Ilvesdal in 1216, this name comes from Old Scandinavian. While this name certainly comes from *ulfr dalr* it is difficult to see if this is 'the valley frequented by wolves' or whether the first element is actually a personal name and thus giving 'Ulfr's valley'.

Ullswater

Records of this name begin with Ulueswater around 1230. Here a Scandinavian personal name and Old English *waeter* tell us this was known as 'the lake of a man called Ulfr'.

Ulpha

A name found as Wolfhou in 1279 shows this comes from Old Scandinavian *ulfr huagr* and describes 'the hill frequented by wolves'.

Ulverston

Domesday records this name as Ulurestun in 1086. The name features a personal name and Old English *tun*. If the man was a Saxon this would have been 'farmstead of a man called Wulfhere', had he been a Scandinavian it would tell of 'Ulfarr's farmstead'.

Underbarrow

The earliest known record is quite late, dating from 1517 as Underbarroe. This is a name of Old English origins in *under beorg* which means the '(place) under the hill'.

Unthank

The earliest record of this name gives the same form as today, seen in 1254. This is an Old English *unthanc* which refers to 'the land held without consent', showing this was held by squatters.

Urswick (Great & Little)

The additions here are self-explanatory and show the smaller version is an off-shoot of the larger and original settlement. Recorded as Ursewica around the middle of the twelfth century, this comes from Old English *ur sae wic* which speaks of 'the specialised or dairy farm by the bison lake'. There is no chance that this name refers to actual bison, these were extinct in our islands three thousand years ago, hence the name probably refers to some natural feature which resembled the animal.

Ushaw Moor

A place name derived from Old English *wulf scaga* and meaning 'the small wood frequented by wolves'. This name is recorded as Ulveskake in the twelfth century.

W

Wackerfield

Listed as Wacarfeld around the middle of the eleventh century, this name comes from Old English *wacor feld* and refers to 'the open land where osiers grow'.

More recent examples of the Sun Inn will suggest a favourable position, earlier this could be heraldic or even chosen for being a simple and easily recognised image.

Waldridge

From Old English *wall brycg* and meaning 'the ridge with or by a wall'. The name is recorded as Walrigge in 1297.

Walworth

Recorded as Walewrth in 1207, this name comes from Old English *walh worth* and tells us this was 'the enclosure of the Welshmen'. The Saxon term *walh* was originally used to describe 'foreigners' and, while it may seem strange today, the vast majority who were not Saxon (thus foreign) were the earlier Celts or Britons.

Washington

Listed as Pipe Wessington in 1197, as Wassinton in 1211, and as Wassyngtona in 1280. Here the temptation to suggest 'the farmstead near the washing place (for sheep)' is tempting, however early forms show this unlikely. Here we probably have 'the farmstead of the family or followers of a man called Wassa'.

Locally we find Barmston, a name from Old English *tun* and a Saxon personal name and speaking of 'the farmstead of a man called Beorn'. Biddick unites Old Scandinavian *by* and Old English *dic* and refers to 'the village with a trench'. Swinhope can still be seen as 'the valley where swine are kept'. Usworth is first seen as Useworth, Osseworth, Oseworth and Usseworth, this appears to be 'the enclosure of a man called Osa'. Cresswell describes 'the spring or stream where watercress grows'.

Wear (River)
Listed as Ufuri in 750 and as Wyry in 1200, this is a Celtic river name and thus one of highly simplistic meaning. Here the name describes 'the winding stream'.

West Auckland
As with other Aucklands nearby this tells us of 'the rock or hill on a river called Clyde', the early name of the River Gaunless. Listed as Alclit around 1040, the additional West shows the other two are to the east.

Luterington is a local place name derived from 'the farmstead associated with a man called Hlothere'. Bildershaw is from Old English *sceaga* with a personal name and tells of 'the woodland enclosure of a man called Bilheard or Bilkere'. Widehope describes the 'wider valley'.

West Rainton
A place name first found around 1170, where it is seen as Reiningtone. Here the Old English *ing tun* follows a Saxon personal name and speaks of 'the farmstead associated with a man called Raegen or Regna'. The addition is to distinguish this place from similarly named places in neighbouring North Yorkshire.

Westwick
Recorded as Westewic in 1091, this name comes from the Old English *west wic* and speaks of 'the specialised farm to the west'. As with most examples, that speciality was almost certainly dairy produce.

Whickham
Recorded as Quicham in 1196, this name probably comes from Old English *cwic hege* and describes 'the homestead with a quickset hedge'.

Byermoor is found as Becchermore in 1385, this being from 'the moor of a man called Beaghere'. The first element of Farnacres is a northeast dialect term *farn* with Old English *aecer* and describing 'the agricutural lands marked by ferns'. Hollingside, seen as Holynsyde in 1382, is probably from *holegn* 'the slope where holly grows'. Swalwell is a Saxon place name describing 'the springs or wells where swallows are seen'. Whaggs is an unusual form of *quag* meaning 'the bogs'.

Whitburn

There is no doubting this comes from Old English *hwit burna* and describes 'the white spring or stream'.

Locally we find The Bents, a place name which must come from Old English *beonet* meaning 'where bent grass grows'. Clearly grass does not grow bent, however it will collapse under its own weight, especially during heavy rain or strong winds. From this we can deduce this area was not ploughed, nor was it grazed by either domesticated or wild animals.

Whorlton

The earliest surviving record comes from the middle of the eleventh century where it appears as Queorningtun. The name comes from Old English *cweorn ing tun* and describes 'the farmstead at the place associated with mill stones'.

Minor names here include Sledwick, recorded as Sliddewisse and Slideuesse this name probably represents Old English *slaed wase* or 'the boggy valley'. Stubb House is built on 'the area marked by stubbs or tree stumps'.

Willington

Seen as Wyvelintun in 1190, this name comes from Old English, where the common *tun* with *ing* and follows a Saxon personal name and tells us this was 'the farmstead of the family or followers of a man called Wifel'.

The Lion and Unicorn is a reminder of the location near the border with Scotland, the Lion representative of England, the Unicorn stands for Scotland.

Wingate

A name recorded in 1071 as Windegatum, this comes from Old English *wind geat* and describes 'the exposed or windy gap or pass'.

Embleton is a minor place name which probably comes from *emel dun*, this is Old English for 'the hill infested by caterpillars'. Although a personal name such as Aemele cannot be ruled out.

Two main lines converged on Wingate during the height of the railways. Although no track survives a gate from a level crossing is a reminder of the origins of the Crossings public house. The Corner House reveals its location and uses this as a marker, as does the Fir Tree Inn.

Willington's floral addition makes all the difference to the sign

Winlaton

Listed as Winloctune in 1085, and as Winlaketon in the twelfth century, this features Old English *tun* and a Saxon personal name and tells of 'the farmstead of a man called Winelac'.

Here also is the hamlet of Chopwell, found as Chapwell in 1416 this is 'the hill or mound of a man called Ceapa', with the Saxon personal name followed by Old English *hlaw*.

Winston

A name which comes from Old English *tun* and a Saxon personal name and describing 'the farmstead of a man called Wine'. The earliest known record of this name dates from 1091 as Winestona.

Local names include Barford, a fairly common Saxon place name and one seen as coming from Old English *beorht ford* or 'the bright ford'. However it should also be noted this first element could also be used as a Saxon personal name. Osmond Croft has not changed since the early fourteenth century, it still describes 'the croft or smallholding of a man called Osmund'.

The Bridgewater Arms comes from the high bridge arching over the nearby Tees. In 1988 a stuntman flew a Spitfire under the bridge at 200mph for a scene in the series *A Piece of Cake*.

Witton Gilbert

The first record of this name comes from 1195 where it appears as Wyton. The basic name here is from Old English *wudu tun* and describes 'the farmstead by the wood'. The addition is manorial, referring to it being held by Gilbert de la Ley by the twelfth century.

Local names include Maidenstonhall, found as Maydencastell, Maidestane, Maydenstanhall and Maidenstane from the middle of the fourteenth century. Clearly this was once described as a 'castle', however this is probably due to confusion with Maiden Castle. Most likely this is from a Saxon personal name with Old English *stan halh* and giving 'the nook of land at the boundary stone of a man called Madan'. Fulford is a common enough name for what would have been a regularly encountered 'muddy ford'. Earlshouse can only be describing 'the home of a man called Earle'.

The White Tun is a play on words, the sign featured a rebus, a pictogram where pictures represent syllables. Here 'the white tun', a large cask for beers and wines, represents the place name of Witton.

Witton-le-Wear

Another place name derived from *wudu tun*, indeed the name is recorded as Wudutun during the eleventh century, again describing 'the farmstead by the wood'. Here the addition refers to the River Wear, a Celtic river name meaning simply 'water' which is discussed under its own entry.

Edge Knoll is not quite self-explanatory, early forms such as

An array of Durham place names on this sign at Witton-le-Wear

135

Edenes knoll in 1303 shows this to be 'the hill of a man called Edwin'. Linburn Beck is from *lin burna* 'the pool with a beck or stream'.

Fitches is recorded as Fychewache in 1382 and as Fyccheworth in 1392, this is difficult to define but may represent *fiche worth* or 'the enclosure where vetch grows'. The vetches were among the first domesticated crops, grown in the Near East around 9,500 years ago. With passing centuries the crop became less popular for human consumption, although during times of starvation vetches 'featured in the frugal diet of the poor' as late as the eighteenth century and were even sold on the black market in southern France during the German occupation of the Second World War.

The Victoria Inn is one of the many pubs named after our longest reigning monarch (1837-1901), no woman has more public houses named after her.

Church of St Philip and St James in Witton-le-Wear

The Victoria public house, Witton-le-Wear

Wolsingham

Recorded in the middle of the twelfth century as Wisingham, this name comes from a Saxon personal name and Old English *inga ham* and telling us of 'the homestead of the family or followers of a man called Wulfsige'.

To the north is Wisserley, this name comes from *wassen leah* and describes 'the wet low-lying woodland clearing'. Broadwood holds no surprises, first recorded in 1153 it does indeed mean 'the broad wood' but also tells us the alignment in relation to the main road, if this was along the same line as the way it would be described as 'the long wood'. Pawlaw Pike is either 'the *feld* or open land where pagan rituals are held' or there is a small chance the first element is the personal name Paga.

Redmires may seem to describe 'the red mire' but logically 'the reedy swamp' makes much more sense. Fawnlees is thought to be a dialect term, *faugh leah* describing 'the variegated woodland clearing', a reference to the colour of the land not the flora. Goosecroft holds no surprises, this being 'the small holding where geese are reared'. Greenwell is a reminder that this 'spring or stream is overgrown'.

Harelaw is from *har hlaw* or 'the mound or tumulus marked by a stone or stones'. Heatherley Clough is seen as Hethereclogh in 1423, this is probably 'the *cloh* or ravine of a man called Haethhere' rather than 'the ravine where heather grows'. Ladley is an Old English name describing 'the *leah* or woodland clearing of a man called Ladda'.

It is not difficult to see Landieu as coming from an Anglicised French term for, literally, 'the land associated with God' but speaking of agricultural land under the control of the church. There seems little doubt White Kirkley describes 'the white kirtle land', although this has never been understood. Wigside is recorded as Wygesyde 1382 and refers to 'the hill of a man called Wicga'.

Wolviston
A name found as Oluestona in 1091 and as Wluestuna in the twelfth century. Here the name describes 'the farmstead of a man called Uluf'.

Locally we find Hannagate Cottage, a name speaking of 'the way to the gate or gap'; Haydike Lane was named for being 'the enclosure with a ditch'; and the Hurle is from *hwerfel* meaning 'a circle' and a likely reference to a stone circle or similar feature. Wilmire House takes the name which began as 'the pool of a man called Wifel'.

Wycliffe
Domesday records this name as Witcliue in 1086, this comes from Old English *hwit clif* and refers to 'the white cliff or bank'.

Wynyard

The oldest surviving record of this name comes from 1208 as Winyard, a name from old English *winn geard* and describing 'the meadow enclosure'.

Locally we find Burntoft, a combination of Old English *berned* and Old Scandinavian *toft* which describes 'the homestead in the area cleared by burning'. Swainston has changed little since the fourteenth century listing of Swayneston, this describing 'the farmstead of a man called Sveinn'.

Common place-name elements

Element	Origin	Meaning
ac	Old English	oak tree
banke	Old Scandinavian	bank, hill slope
bearu	Old English	grove, wood
bekkr	Old Scandinavian	stream
berg	Old Scandinavian	hill
birce	Old English	birch tree
brad	Old English	broad
broc	Old English	brook, stream
brycg	Old English	bridge
burh	Old English	fortified place
burna	Old English	stream
by	Old Scandinavian	farmstead
ceap	Old English	market
ceaster	Old English	Roman stronghold
cirice	Old English	church
clif	Old English	cliff, slope
cocc	Old English	woodcock
cot	Old English	cottage
cumb	Old English	valley
cweorn	Old English	queorn
cyning	Old English	king
dael	Old English	valley
dalr	Old Scandinavian	valley
denu	Old English	valley
draeg	Old English	portage
dun	Old English	hill
ea	Old English	river
east	Old English	east
ecg	Old English	edge
eg	Old English	island
eorl	Old English	nobleman
eowestre	Old English	fold for sheep

fald	Old English	animal enclosure
feld	Old English	open land
ford	Old English	river crossing
ful	Old English	foul, dirty
geard	Old English	yard
geat	Old English	gap, pass
haeg	Old English	enclosure
haeth	Old English	heath
haga	Old English	hedged enclosure
halh	Old English	nook of land
ham	Old English	homestead
hamm	Old English	river meadow
heah	Old English	high, chief
hlaw	Old English	tumulus, mound
hoh	Old English	hill spur
hop	Old English	enclosed valley
hrycg	Old English	ridge
hwaete	Old English	wheat
hwit	Old English	white
hyll	Old English	hill
lacu	Old English	stream, water course
lang	Old English	long
langr	Old Scandinavian	long
leah	Old English	woodland clearing
lytel	Old English	little
meos	Old English	moss
mere	Old English	lake
middel	Old English	middle
mor	Old English	moorland
myln	Old English	mill
niwe	Old English	new
north	Old English	north
ofer	Old English	bank, ridge
pol	Old English	pool, pond
preost	Old English	priest
ruh	Old English	rough
salh	Old English	willow
sceaga	Old English	small wood, copse
sceap	Old English	sheep
stan	Old English	stone, boundary stone

steinn	Old Scandinavian	stone, boundary stone
stapol	Old English	post, pillar
stoc	Old English	secondary or special settlement
stocc	Old English	stump, log
stow	Old English	assembly or holy place
straet	Old English	Roman road
suth	Old English	south
thorp	Old Scandinavian	outlying farmstead
treow	Old English	tree, post
tun	Old English	farmstead
wald	Old English	woodland, forest
wella	Old English	spring, stream
west	Old English	west
wic	Old English	specialised, usually dairy arm
withig	Old English	willow tree
worth	Old English	an enclosure
wudu	Old English	wood

Bibliography

Dunkling, Leslie and Wright, Gordon *A Dictionary of Pub Names*
Ekwall, Eilert *The Concise Oxford Dictionary of English Place-Names*
Jackson, Rev Charles E. *The Place Names of Durham* by
Mawer, A. *The Place Names of Northumberland and Durham*
Mills, A.D. *Oxford Dictionary of English Place Names*

This Is Gateshead
Durham Times
Sunderland Street Names
Harry Bilton and Newton Aycliffe Library

Also from Sigma Leisure:

Northumberland Place Names
Anthony Poulton-Smith

Ever wondered why our towns and villages are named as they are? Who named them and why? Did corn grow on the hill at Cornhill-on-Tweed? How was Haltwhistle named such before the railways arrived? Was Ogle named as a good look-out point? And did a snake give a name Adderstone?

Within these pages author Anthony Poulton-Smith examines the origins of the place names with which we are otherwise so familiar. Towns, villages, districts, hills, streams, woods, farms, fields, streets and even pubs are examined and explained. Some of the definitions give a glimpse of life in the earlier days of the settlement, and for the author there is nothing more satisfying than finding a name which gives such a snapshot. The definitions are supported by anecdotal evidence, bring to life the individuals and events which have influenced the places and how these names have developed. This is not simply a dictionary but a history and will prove invaluable not only for those who live and work in the county but also visitors and tourists, historians and former inhabitants, indeed anyone with an interest in Northumberland and the city of Newcastle upon Tyne.

£8.99

Cambridgeshire Place Names
Anthony Poulton-Smith

Ever wondered why our towns and villages are named as they are? Who named them and why? Within these pages author Anthony Poulton-Smith examines the origins of the place names with which we are otherwise so familiar. Towns, villages, districts, hills, streams, woods, farms, fields, streets and even pubs are examined and explained. Some of the definitions give a glimpse of life in the earlier days of the settlement, and for the author there is nothing more satisfying than finding a name which gives such a snapshot. The definitions are supported by anecdotal evidence, bring to life the individuals and events which have influenced the places and the way these names have developed.

This is not simply a dictionary but a history and will prove invaluable not only for those who live and work in the county but also visitors and tourists, historians and former inhabitants, indeed anyone with an interest in Cambridgeshire.

£9.99

All of our books are available through booksellers.
For a free catalogue, please contact:
Sigma Leisure, Stobart House, Pontyclerc
Penybanc Road, Ammanford SA18 3HP
Tel: 01269 593100 Fax: 01269 596116

info@sigmapress.co.uk   www.sigmapress.co.uk